WOODLAND LIFE

By G. MANDAHL-BARTH

Edited in the English Edition

By ARNOLD DARLINGTON

Colour plates by
HENNING ANTHON

BLANDFORD PRESS

LONDON

First published in this edition 1966
2nd edition 1972
Reprinted 1974
© Copyright 1966 by Blandford Press Ltd.,
167 High Holborn, London, W.C.1.

ISBN 0 7137 0417 9

Originally published in Denmark as
HVAD FINDER JEG I SKOVEN
by Politikens Forlag 1966

Translated from the Danish by
Edwin Homstrom, Ph.D., F.I.L.

Colour printed in Denmark by
F. E. Bording A/S, Copenhagen
Text printed in Great Britain by
Fletcher & Son Ltd, Norwich

PREFACE TO THE ENGLISH EDITION

The great majority of organisms listed in the Danish edition also occur in the British Isles, where many of them are widely distributed and, in places, abundant. The few not recognised as British have been retained in the present edition for two reasons. Firstly, most of them are so similar to species undoubtedly present in these islands that their particulars are useful when attempting broad identifications. Secondly, there is always the possibility that some of them have been overlooked, or that they may put in an appearance from time to time. Clearly, smaller animals are more likely to be missed than larger ones. Among the bigger invertebrates whose status is uncertain is the pine-tree lappet moth (*Dendrolimus pini* – 148, 219) which, although probably widespread in Denmark, has been reliably identified on few occasions in Britain.

Useful references for further reading include the following works in English:

Beirne B. P. (1954) *British Pyralid and Plume Moths*. Warne.

Cloudsley-Thompson, J. L. and Sankey J. (1961) *Land Invertebrates*. Methuen.

Collyer, C. N. and Hammond, C. O. (1951) *Flies of the British Isles*. Warne.

Kevan, D. K. McE. (1962) *Soil Animals*. Witherby.

Linssen, E. F. (1959) *Beetles of the British Isles*. Warne.

Locket, G. H. and Millidge, A. F. (1953) *British Spiders*. Ray Society.

Sandars, E. (1946) *An Insect Book for the Pocket*. Oxford U.P.

South, R. (1928) *The Butterflies of the British Isles*. Warne.

South, R. (1943) *The Moths of the British Isles*. Warne.

Southwood, T. R. E. and Leston, D. (1959) *Land and Water Bugs of the British Isles*. Warne.

Step, E (1932) *Bees, Wasps, Ants and Allied Insects of the British Isles*. Warne.

Arnold Darlington,
Malvern College.

FOREWORD

This book has been written for those who like to visit woodlands, to observe the wild life which is found there. It should be said at the outset that it does not cover everything. For instance, it includes none of the vertebrates, such as birds and small mammals, which abound in timbered areas. It deals with a selection of the invertebrates (worms, woodlice, insects, spiders, snails and so on), and in many instances gives examples of their relationships with one another and with their surroundings. The term 'woods' is taken in a wide sense, so that reference is also made to the fauna found in timbered gardens, scrubland, hedgerows and trees lining roadsides. Examples of the main groups of invertebrates colonising woods, which can be distinguished either by their size or the signs of their presence, can be found in the book. Of the small, almost invisible types – and these are the most numerous – a few representative species have been selected, which should enable the reader to place the majority of the woodland creatures in the right categories.

The book has been divided into two parts. The first and larger section comprises the coloured plates, pages 5-85, and presents the animals in systematic order. With regard to the grouping of the larger insects, the adults are given first, followed by separate plates showing the larvae of some species. Not all the organisms are drawn to the same scale; but the actual size is generally indicated in the text, either as length of head and body or wing span. The smaller section consists of 15 pages illustrating some of the ways in which insects and mites can damage plants. The figures in these are nearly all drawn full size and are in botanical order. Thus if one finds a gall it should be possible to discover immediately whether it was made by a mite, a fly or a wasp; as well as the species of plant which is involved.

In instances where an organism has an accepted popular name, this has been given in the plates and in the text along with the scientific name. Many have no recognised popular name.

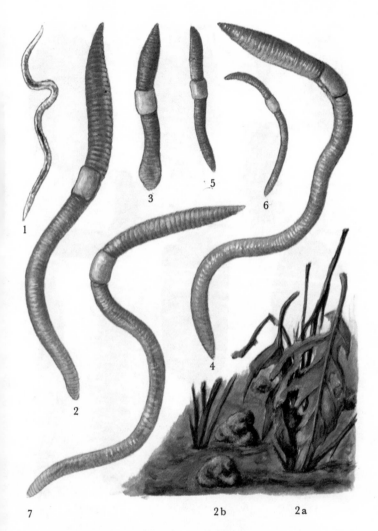

OLIGOCHAETA EARTH WORMS

1 **Pot Worm,** *Mesenchytraeus setosus* 2 **Marsh Worm,** *Lumbricus rubellus*
2a Partly buried leaves 2b Faeces 3 *L. castaneus* 4 *Allolobophora turgida*
5 *Dendrobaena octaedra* 6 *D. arborea* 7 *Octolasium cyaneum*

6

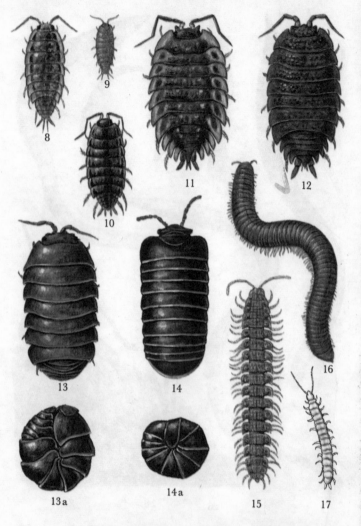

CRUSTACEA: WOODLICE, AND DIPLOPODA: MILLEPEDES

8 **Woodlouse,** *Ligidium hypnorum* 9 **Woodlouse,** *Trichoniscus pusillus* 10 **Woodlouse,** *Philoscia muscorum* 11 **Woodlouse,** *Oniscus asellus* 12 **Woodlouse,** *Porcellio scaber* 13 **Pill Woodlouse,** *Armadillidium cinereum* 13a contracted 14 **Pill Millepede,** *Glomeris marginata* 14a contracted 15 **Flat-back Millepede,** *Polydesmus complanatus* 16 **Woodland Snake Millepede,** *Cylindroiulus sylvarum* 17 **Dwarf Millepede,** *Scutigerella immaculata*

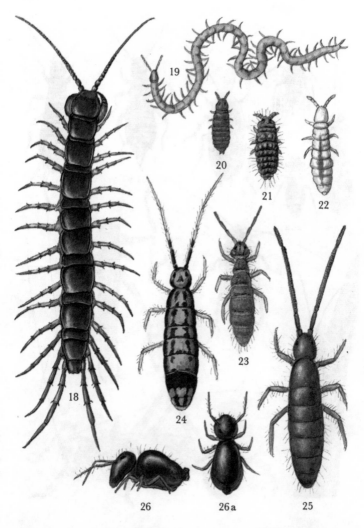

CHILOPODA: CENTIPEDES, AND COLLEMBOLA: SPRINGTAILS
18 **Lithobiomorph Centipede,** *Lithobius forficatus* 19 **Geophilomorph Centipede,** *Geophilus sp.* 20 **Springtail,** *Hypogastrura armata* 21 **Springtail,** *Neanura muscorum* 22 **Springtail,** *Onychiurus armatus* 23 **Springtail,** *Isotoma viridis* 24 **Springtail,** *Orchesella flavescens* 25 **Springtail,** *Tomocerus plumbeus* 26 **Springtail,** *Allacma fusca*

8

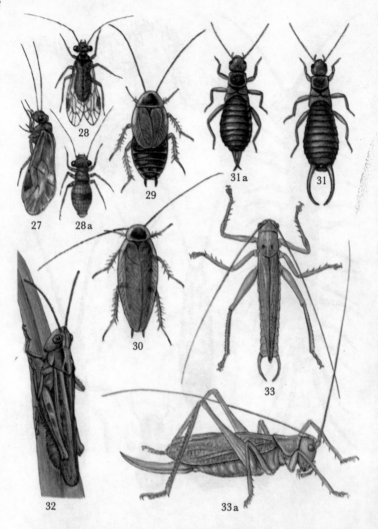

PSOCOPTERA: BOOK LICE, ORTHOPTERA: COCKROACHES AND DERMAPTERA:
GRASSHOPPERS

27 **Book Louse,** *Psocus nebulosus* 28 **Book Louse,** *Mesopsocus unipunctatis,* male
28a female 29 **Cockroach,** *Ectobius lapponicus* 30 **Wood Cockroach,** *E. lividus* 31 **Wood Earwig,** *Chelidurella acanthopygia* 32 **Field Grasshopper,** *Chorthippus* sp. 33 **Bush Cricket,** *Meconema varium*

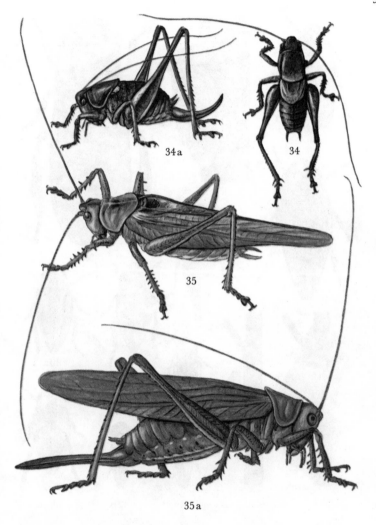

ORTHOPTERA: GRASSHOPPERS
34 **Bush Cricket,** *Pholidoptera griseoaptera,* male 34a female 35 **Great Green Grasshopper,** *Tettigonia viridissima,* male 35a female

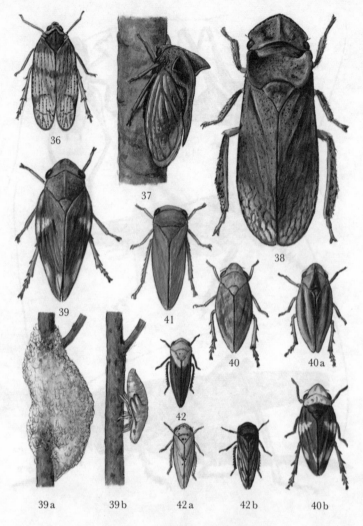

HEMIPTERA – HOMOPTERA: PLANT BUGS

36 *Cixius nervosus* 37 *Centrotus cornutus* 38 *Ledra aurita* 39 *Aphrophora salicis* 39a cuckoo-spit 39b nymph 40 *Philaenus spumarius* 40a and 40b Colour variants 41 *Jassus lanio* 42 *Bythoscopus flavicollis* 42a and 42b Colour variants

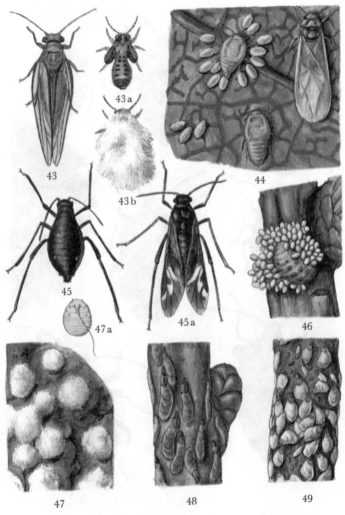

HEMIPTERA – HOMOPTERA: PSYLLDS, APHIDS AND COCCIDS
43 **Jumping Plant Louse,** *Psylla alni* 43a nymph 43b Nymph covered by
'wool' 44 **Oakleaf Aphid,** *Phylloxera quercus*, eggs, nymph, wingless
female, female with wings 45 **Aphid,** *Lachnus exsiccator*, wingless female 45a
female with wings 46 **Aphid,** *Adelges abietis*, female with eggs 47 **Coccid,**
Cryptococcus fagi 47a Underside of female 48 **Coccid,** *Lepidosaphes ulmi* 49
Coccid, *Chionaspis salicis*, males, females and larvae

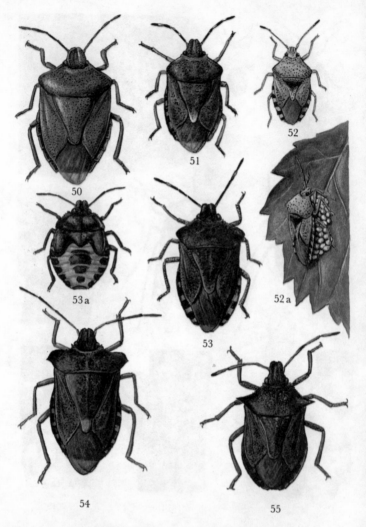

HEMIPTERA–HETEROPTERA : SHIELD-BUGS
50 **Green Shield-bug,** *Palomena prasina* 51 **Sloebug,** *Dolycoris baccarum* 52
Birch Bug, *Elasmucha grisea* 52a female with eggs 53 *Troilus luridus* 53a
Nymph 54 **Common Shield-bug,** *Pentatoma rufipes* 55 *Picromerus bidens*

13

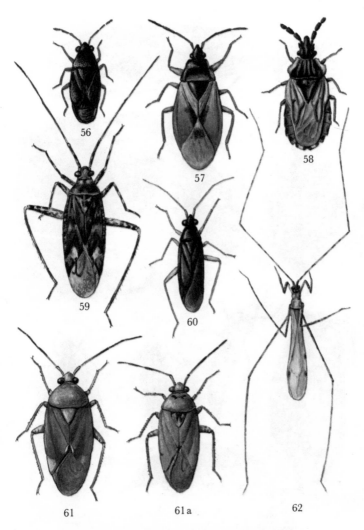

HEMIPTERA – HETEROPTERA: CAPSID AND ASSASSIN BUGS
56 **Capsid Bug,** *Stygnocoris rusticus* 57 **Capsid Bug,** *Gastrodes abietum* 58
Capsid Bug, *Aradus depressus* 59 **Capsid Bug,** *Phytocoris populi* 60 **Capsid Bug,** *Phylus coryli* 61 **Capsid Bug,** *Lygus pratensis,* 'northern' form 61a
'western' form 62 **Assassin Bug,** *Empicoris vagabundus*

14

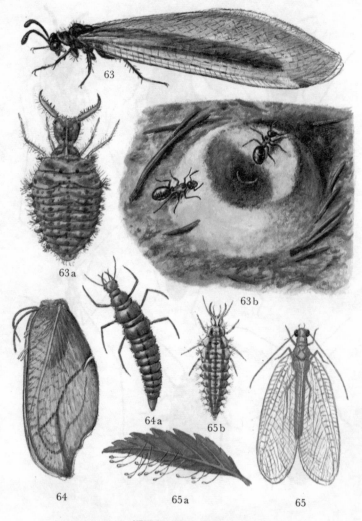

NEUROPTERA : LACEWINGS

63 **Ant-lion,** *Myrmeleon formicarius* 63a larva 63b funnel-shaped trap 64
Dead-leaf Lacewing, *Drepanopteryx phalaenoides* 64a larva 65 **Common
Goldeneye Lacewing,** *Chrysopa vulgaris* 65a eggs 65b larva

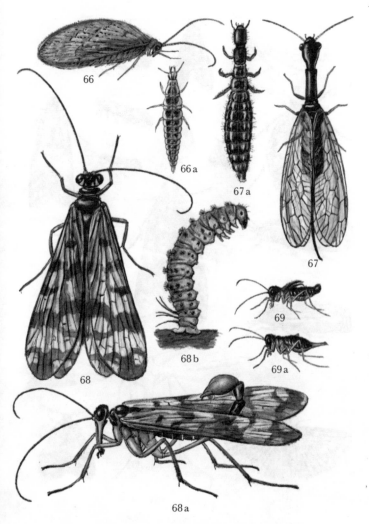

15

NEUROPTERA: LACEWINGS, SNAKE-FLIES AND SCORPION FLIES

66 **Hop Lacewing,** *Hemerobius humuli* 66a larva 67 **Snake-fly,** *Rhaphidia xanthostigma* 67a larva 68 **Common Scorpion Fly,** *Panorpa communis*, female 68a male 68b larva 69 **Snow Flea,** *Boreus hyemalis*, male 69a female

16

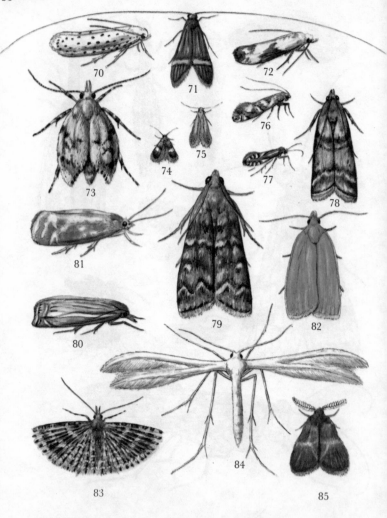

MICROLEPIDOPTERA: 'SMALL' MOTHS

70 **Small Ermine Moth,** *Hyponomeuta euonymella* 71 **Long-horned Moth,**
Adela degeeriella 72 **Ash Moth,** *Prays curticellus* 73 *Chimabache fagella* 74
Nepticula basalella 75 *Tischeria complanella* 76 *Gracilaria syringella* 77 *Lithocolletis faginella* 78 *Acrobasis consociella* 79 *Dioryctria abietella* 80 *Crambus
pratellus* 81 *Evertia buoliana* 82 **Green Tortrix,** *Tortrix viridana* 83
Many-plume Moth, *Orneodes hexadactyla* 84 **White-feather Moth,**
Alucita pentadactyla 85 **Bagworm,** *Fumea casta*

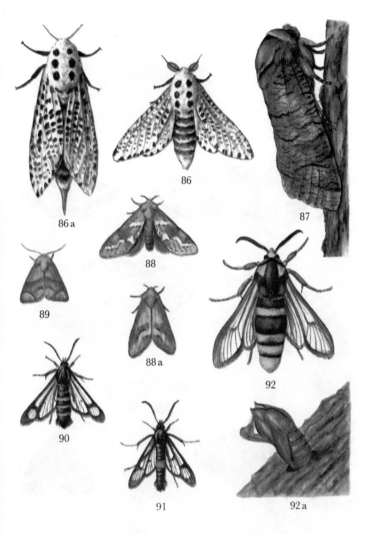

17

LEPIDOPTERA: COSSIDS, CLEARWINGS, SWIFTS AND LIMACODIDS
86 **Leopard Moth,** *Zeuzera pyrina,* male   86a female   87 **Goat Moth,** *Cossus cossus*   88 **Gold Swift,** *Hepialus hecta,* male   88a female   89 **Festoon,** *Apoda avellana*   90 **Raspberry Clearwing,** *Bembecia hylaeiformis*   91 **Large Red-belted Clearwing,** *Aegeria culiciformis*   92 **Hornet Moth,** *Sesia apiformis* 92a Chrysalis after emergence

LEPIDOPTERA: GEOMETRID MOTHS
93 **Chimney-Sweeper,** *Odezia atrata* 94 **Large Emerald,** *Hipparchus papilionaria* 95 **Clouded Magpie,** *Abraxas sylvata* 96 **Clouded Border,** *Lomaspilis marginata* 97 **November Moth,** *Oporinia dilutata* 98 **Light Emerald,** *Campaea margaritata* 99 **Single Dotted Wave,** *Sterrha dimidiata* 100 **Scallop Shell,** *Calocalpe undulata* 101 **Winter Moth,** *Operophtera brumata*, male 101a female

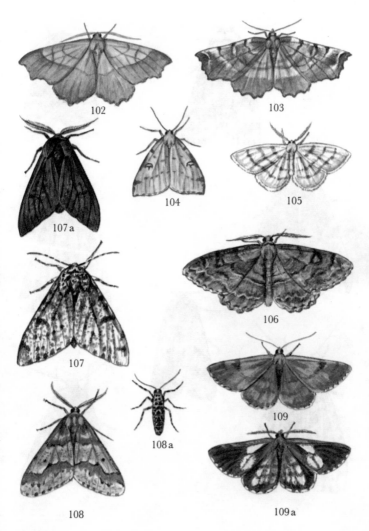

LEPIDOPTERA: GEOMETRID MOTHS

102 **August Thorn,** *Ennomos quercinaria*   103 **Early Thorn,** *Selenia bilunaria*
104 **Brimstone,** *Opisthograptis luteolata*   105 **Common White Wave,**
Cabera pusaria   106 **Mottled Beauty,** *Boarmia repandata*   107 **Peppered
Moth,** *Biston betularia*   107a melanistic form   108 **Mottled Umber,**
Erannis defoliaria, male   108a female   109 **Bordered White,** *Bupalus
piniaria,* male   109a female

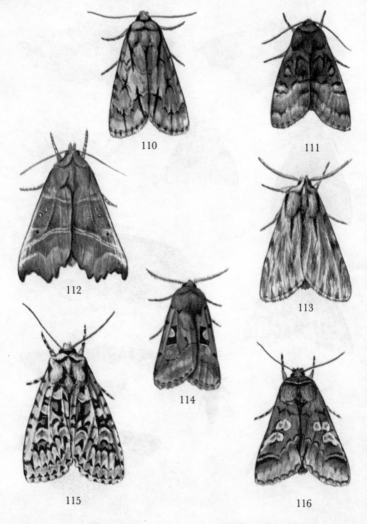

LEPIDOPTERA: NOCTUID AND LYMANTRID MOTHS
110 **Grey Dagger,** *Apatele psi* 111 **Nut-tree Tussock,** *Colocasia coryli*
112 **Herald,** *Scoliopteryx libatrix* 113 **Sprawler,** *Brachionycha sphinx* 114
Hebrew Character, *Orthosia gothica* 115 **Merveille du Jour,** *Griposia aprilina* 116 **Figure of Eight,** *Episema caeruleocephala*

117

118

119

120

LEPIDOTERA: NOCTUID MOTHS
117 Copper Underwing, *Amiphipyra pyramidea*   **118 Large Yellow Under-wing,** *Tryphaena pronuba*   **119 Red Underwing,** *Catocala nupta*   **120 Clifden Nonpareil,** *C. fraxini*

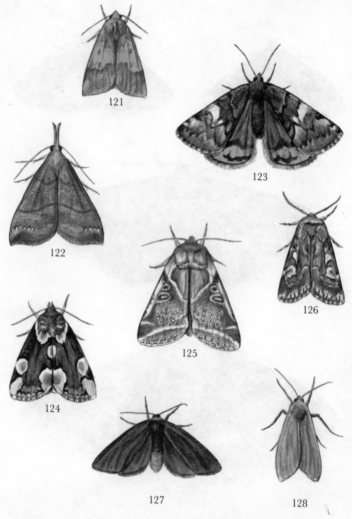

LEPIDOPTERA: NOCTUID, THYATRID AND ARCTIID MOTHS
121 **Barred Sallow**, *Tiliacea aurago* 122 **Snout**, *Hypena proboscidalis* 123
Orange Underwing, *Brephos parthenias* 124 **Peach-blossom**, *Thyatira batis* 125 **Buff Arches**, *Habrosyne derasa* 126 **Pine Beauty**, *Panolis flammea*
127 **Red-necked Footman**, *Atolmis rubricollis* 128 **Common Footman**,
Eilema lurideola

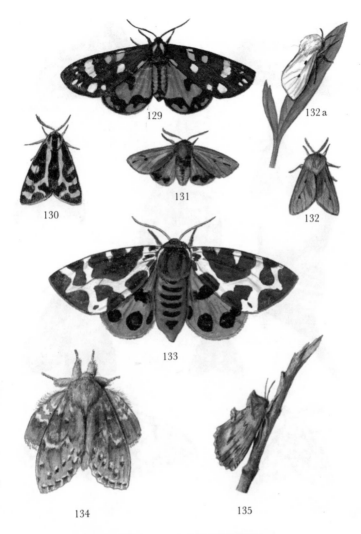

LEPIDOPTERA: TIGERS AND PROMINENTS
129 **Scarlet Tiger,** *Panaxia dominula*   130 **Wood Tiger,** *Parasemia plantaginis*
131 **Ruby Tiger,** *Phragmatobia fuliginosa*   132 **Muslin Moth,** *Diaphora mendica*, male   132a female   133 **Garden Tiger,** *Arctia caja*   134 **Lobster Moth,** *Stauropus fagi*   135 **Coxcomb Prominent,** *Lophopteryx capucina*

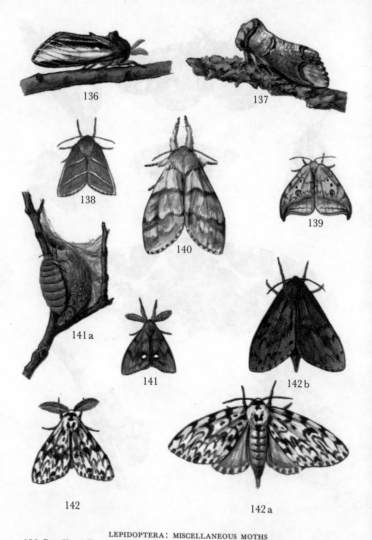

LEPIDOPTERA: MISCELLANEOUS MOTHS

136 **Swallow Prominent,** *Pheosia tremula* 137 **Buff-tip,** *Phalera bucephala*
138 **Green Silver Lines,** *Bena prasinana* 139 **Pebble Hook-tip,** *Drepana falcataria* 140 **Pale Tussock Moth,** *Dasychira pudibunda* 141 **Common Vapourer Moth,** *Orgyia antiqua* male 141a female with eggs on pupa web
142 **Black Arches,** *Lymantria monacha,* male 142a, normal female 142b melanistic female

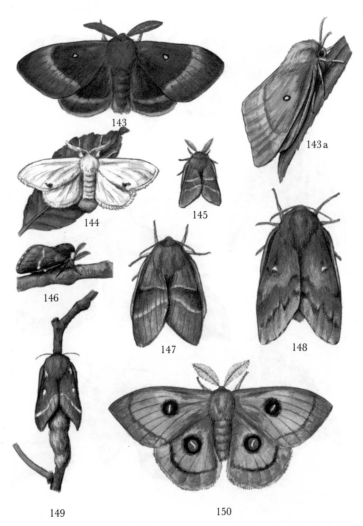

LEPIDOPTERA: MOTHS WITH SPINNING LARVAE

143 **Oak Eggar,** *Lasiocampa quercus,* male 143a female 144 **Yellowtail moth,** *Euproctis similis* 145 **Lackey Moth,** *Malacosoma neustria* 146 **December Moth,** *Poecilocampa populi* 147 **Fox Moth,** *Macrothylacia rubi* 148 **Pine-tree Lappet,** *Dendrolimus pini* 149 **Small Eggar,** *Eriogaster lanestris* 150 **Saturnid,** *Aglia tau*

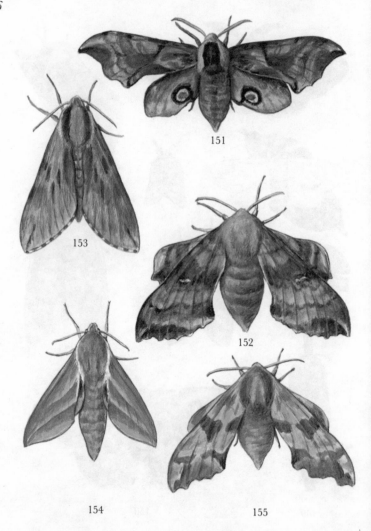

LEPIDOPTERA: NOCTURNAL HAWK MOTHS

151 **Eyed Hawk,** *Smerinthus ocellata*   152 **Poplar Hawk,** *Laothoe populi*
153 **Pine Hawk,** *Hyloicus pinastri*   154 **Elephant Hawk,** *Deilephila elpenor*
155 **Lime Hawk,** *Mimas tiliae*

156

157

158

159

159a

160

LEPIDOPTERA: DIURNAL HAWK MOTHS AND WHITE BUTTERFLIES
156 **Hummingbird Hawk,** *Macroglossum stellatarum*   157 **Broad-bordered
Bee Hawk,** *Hemaris fuciformis*   158 **Black Apollo,** *Parnassius mnemosyne*
159 **Green-veined White,** *Pieris napi*   159a Underside   160 **Brimstone,**
Gonepteryx rhamni

LEPIDOPTERA: NYMPHALID BUTTERFLIES
161 **White Admiral,** *Limenitis sibylla*   161a Underside   162 *Limenitis populi*
163 **Purple Emperor,** *Apatura iris*

164a   164   165   166

LEPIDOPTERA: NYMPHALID BUTTERFLIES
164 **Small Tortoiseshell,** *Aglais urticae* 164a Underside 165 **Peacock,**
Vanessa io 166 **Camberwell Beauty,** *Euvanessa antiopa*

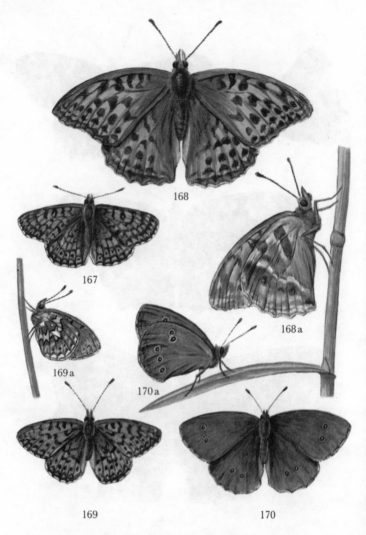

LEPIDOPTERA: NYMPHALID AND SATYRNID BUTTERFLIES
167 **Glanville Fritillary,** *Melitaea cinxia* 168 **Silver-washed Fritillary,**
Argynnis paphia 168a Underside 169 **Pearl-bordered Fritillary,** *A.*
euphrosyne 169a Underside 170 **Ringlet,** *Aphantopus hyperanthus* 170a
Underside

LEPIDOPTERA: SATYRNID, LYCAENID AND PAMPHILINE BUTTERFLIES
171 **Speckled Wood,** *Pararge egeria* 172 **Small Heath,** *Coenonympha pamphilus* 173 **Green Hairstreak,** *Callophrys rubi* 173a Underside 174 **White-letter Hairstreak,** *Thecla w-album* 174a Underside 175 **Purple Hairstreak,** *Zephyrus quercus* 176 **Small Copper,** *Heodes phloeas* 177 **Common Blue,** *Polyommatus icarus* 178 **Essex Skipper,** *Adopaea lineola* 179 **Large Skipper,** *Augiades sylvanus*

LEPIDOPTERA: MISCELLANEOUS MOTH LARVAE
180 *Incurvaria koerneriella* 181 **Small Ermine Moth,** *Hyponomeuta euonymella*
182 **Bagworm,** *Talaeporia tubulosa* 183 **Bagworm,** *Fumea casta* 184
Bagworm, *Coleophora laricinella* 185 **Green Tortrix,** *Tortix viridana* 186
Acrobasis consociella 187 **Goat Moth,** *Cossus cossus* 188 **Festoon,** *Apoda avellana*

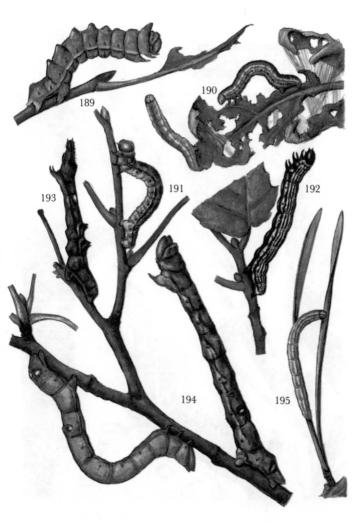

LEPIDOPTERA: GEOMETRID MOTH LARVAE
189 **Large Emerald,** *Hipparchus papilionaria*   190 **Winter Moth,** *Operophtera brumata*   191 **Mottled Umber,** *Erannis defoliaria*   192 **Clouded Magpie,** *Abraxas sylvata*   193 **Early Thorn,** *Selenia bilunaria*   194 **Peppered Moth,** *Biston betularia*, green and brown varieties   195 **Bordered White,** *Bupalus piniaria*

34

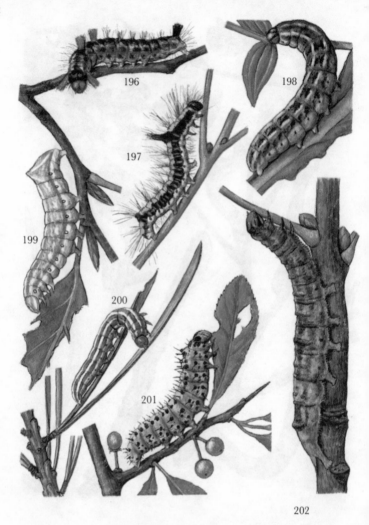

LEPIDOPTERA: NOCTUID MOTH LARVAE

196 **Nut-tree Tussock,** *Colocasia coryli* 197 **Grey Dagger,** *Apatele psi*
198 **Large Yellow Underwing,** *Tryphaena pronuba* 199 **Copper Underwing,** *Amphipyra pyramidea* 200 **Pine Beauty,** *Panolis flammea* 201 **Figure of
Eight,** *Episema caeruleocephala* 202 **Red Underwing,** *Catocala nupta*

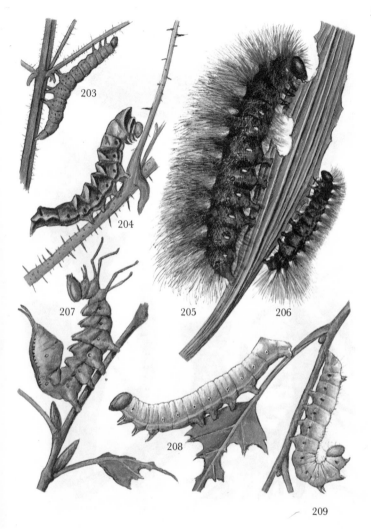

35

LEPIDOPTERA: SPINNING LARVAE OF MOTHS
203 **Snout,** *Hypena proboscidalis* 204 **Peach-blossom,** *Thyatira batis* 205
Garden Tiger, *Arctia caja* 206 **Wood Tiger,** *Parasemia plantaginis* 207
Lobster Moth, *Stauropus fagi* 208 **Swallow Prominent,** *Pheosia tremula*
209 **Coxcomb Prominent,** *Lophopteryx capucina*

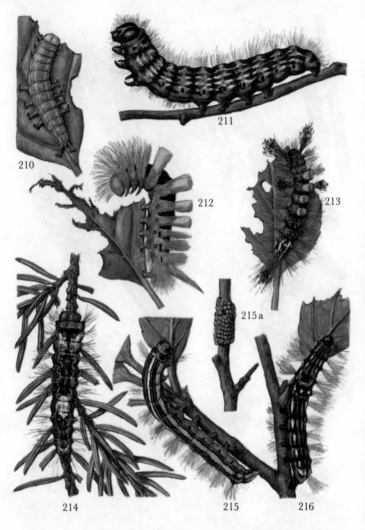

36

LEPIDOPTERA: SPINNING LARVAE OF MOTHS
210 **Green Silver Lines,** *Bena prasinana* 211 **Buff-tip,** *Phalera bucephala*
212 **Pale Tussock,** *Dasychira pudibunda* 213 **Common Vapourer,** *Orgyia
antiqua* 214 **Black Arches,** *Lymantria monacha* 215 **Lackey Moth,**
Malacosoma neustria 215a Eggs 216 **Yellowtail Moth,** *Euproctis similis*

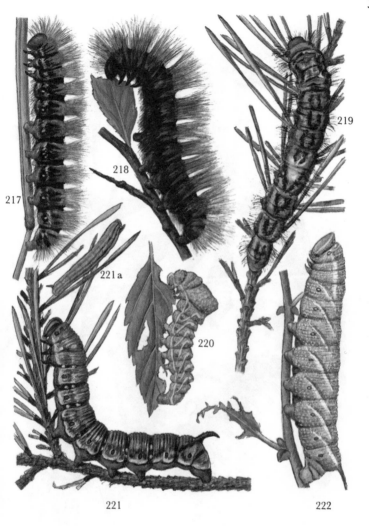

LEPIDOPTERA: LARVAE OF SPINNERS AND HAWK MOTHS
217 **Oak Eggar,** *Lasiocampa quercus* 218 **Fox Moth,** *Macrothylacia rubi*
219 **Pine-tree Lappet,** *Dendrolimus pini* 220 **Saturnid,** *Aglia tau* 221 **Pine Hawk,** *Hyloicus pinastri* 221a Larvae 222 **Eyed Hawk,** *Smerinthus ocellata*

LEPIDOPTERA: LARVAE OF HAWK MOTHS AND BUTTERFLIES
223 **Elephant Hawk,** *Deilephila elpenor* 224 **Green-veined White,** *Pieris napi* 224a pupa 225 **Brimstone,** *Gonepteryx rhamni* 225a pupa 226 **Small Tortoiseshell,** *Aglais urticae* 226a pupa 227 **Peacock,** *Vanessa io* 227a pupa 228 **Pearl-bordered Fritillary,** *Argynnis euphrosyne* 229 **Common Blue,** *Polyommatus icarus*

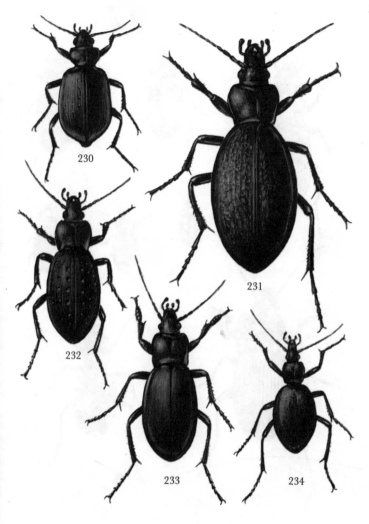

COLEOPTERA: GROUND BEETLES

230 *Calosoma inquisitor*   231 *Carabus coriaceus*   232 *C. hortensis*   233 **Violet
Ground Beetle,** *C. violaceus*   234 *Cychrus caraboides* var. *rostratus*

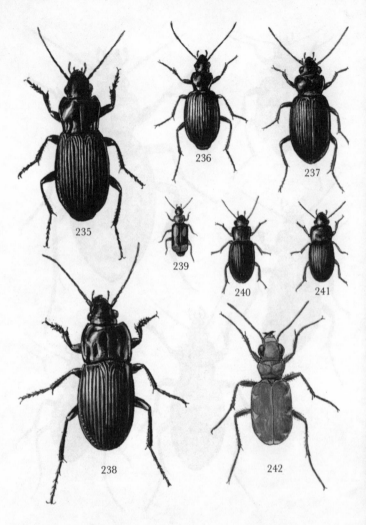

COLEOPTERA: GROUND BEETLES

235 *Feronia niger* 236 *Platynus assimilis* 237 *Nebria brevicollis* 238 *Abax parallelopipedus* 239 *Dromius quadrimaculatus* 240 *Ophonus seladon* 241 *Harpalus latus* 242 **Tiger Beetle**, *Cincindela campestris*

41

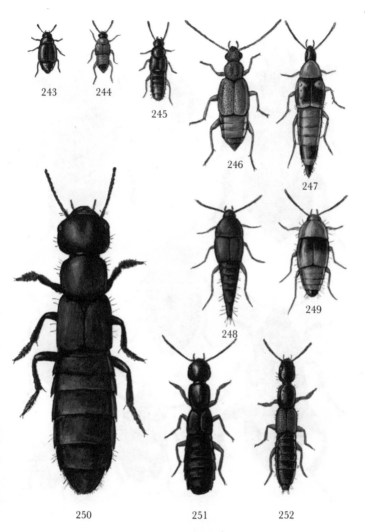

COLEOPTERA: CARRION BEETLES
243 *Proteinus brachypterus* 244 *Gyrophaena affinis* 245 *Atheta fungi* 246
Anthophagus caraboides 247 *Bolitobius lunulatus* 248 *Conosoma testaceum*
249 *Tachyporus obtusus* 250 **Devil's Coach-horse,** *Ocypus olens* 251
Staphylinus brunnipes 252 *Othius punctulatus*

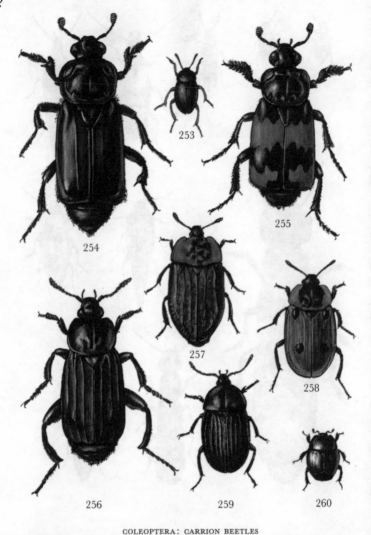

COLEOPTERA: CARRION BEETLES

253 *Catops picipes* 254 **Sexton Beetle,** *Necrophorus humator* 255 **Sexton Beetle,** *N. investigator* 256 **Sexton Beetle,** *Necrodes litoralis* 257 **Red-breasted Carrion Beetle,** *Oeceoptoma thoracica* 258 *Xylodrepa quadripunctata* 259 **Black Carrion Beetle,** *Silpha carinata* 260 *Hister striola*

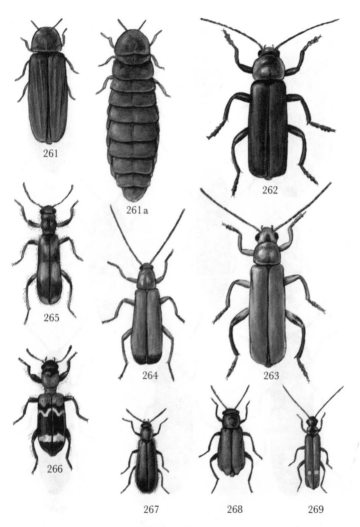

COLEOPTERA: 'SOFTWING' BEETLES

261 **Glow-worm,** *Lampyris noctiluca,* male 261a female 262 **Soldier Beetle,** *Cantharis fusca* 263 **Soldier Beetle,** *C. livida* 264 *Rhagonycha fulva* 265 *Opilo mollis* 266 **Ant Beetle,** *Thanasimus formicarius* 267 *Dasytes caeruleus* 268 *Malachius bipustulatus* 269 *Malthodes marginatus*

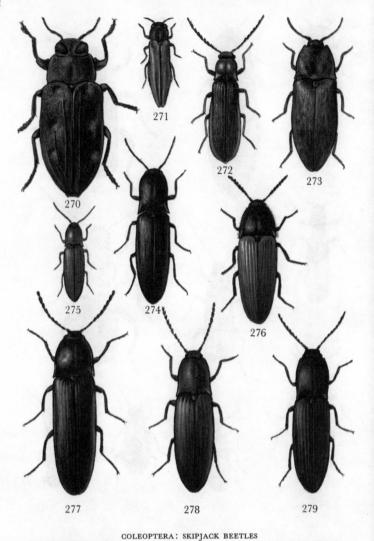

44

COLEOPTERA: SKIPJACK BEETLES
270 *Chrysobothris affinis* 271 *Agrilus viridis* 272 *Denticollis linearis* 273 *Lacon murinus* 274 *Agriotes aterrimus* 275 *A. acuminatus* 276 *Elater cinnabarinus* 277 *Melanotus rufipes* 278 *Corymbites sjaelandicus* 279 *Athous haemorrhoidalis*

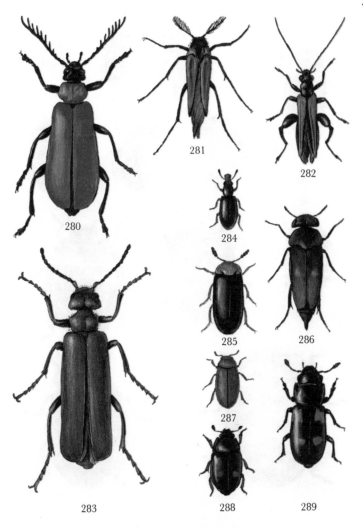

45

COLEOPTERA: MISCELLANEOUS BEETLES

280 **Cardinal Beetle,** *Pyrochroa coccinea*   281 *Metoecus paradoxus*   282 *Oedemera femorata*   283 **Spanish Fly,** *Lytta vesicatoria*   284 *Rhinosimus planirostris* 285 *Tetratoma fungorum*   286 *Tomoxia biguttata*   287 **Raspberry Beetle,** *Byturus urbanus*   288 *Nitidula bipunctata*   289 *Glischrochilus* var. *quadripunctatus*

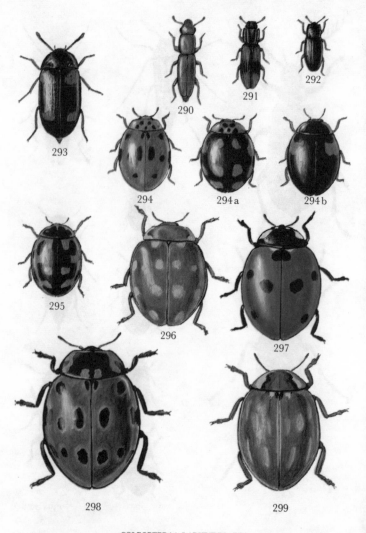

COLEOPTERA: LADYBIRDS, ETC.

290 *Rhizophagus dispar* 291 *Bitoma crenata* 292 *Cerylon histeroides* 293 **Fungus Beetle**, *Mycetophagus quadripustulatis* 294 **Ten-spot Ladybird,** *Coccinella decempunctata* 294a and 294b colour variations 295 **Fourteen-spot Ladybird,** *C. quatuordecimpunctata* 296 *C. quatuordecimguttata* 297 **Seven-spot Ladybird,** *C. septempunctata* 298 **Eyed Ladybird** *Anatis ocellata* 299 *Paramysia oblongoguttata*

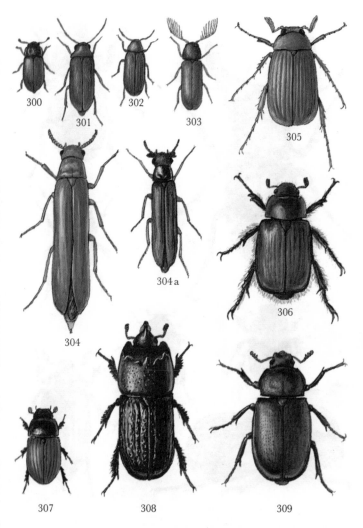

COLEOPTERA: BORING AND SCARABAEID BEETLES
300 **Boletus Beetle,** *Cis boleti* 301 *Ernobius mollis* 302 **Furniture Beetle,** *Anobium punctatum* 303 **Furniture Beetle,** *Ptilinus pectinicornis* 304 *Hylecoetus dermestoides,* female 304a male 305 *Serica brunnea* 306 **Garden Chafer,** *Phyllopertha horticola* 307 *Aphodius fimetarius* 308 *Sinodendron cylindricum* 309 *Systenocerus caraboides*

48

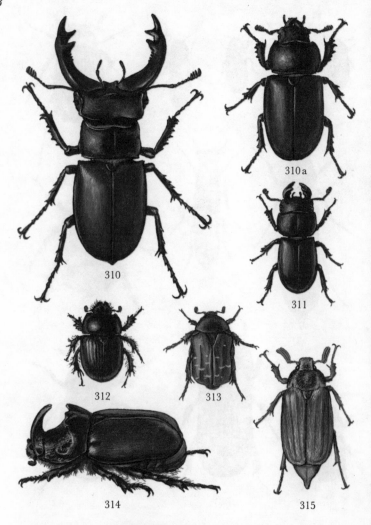

COLEOPTERA: SCARABAEID BEETLES

310 Greater Stag Beetle, *Lucanus cervus,* male   310a female   **311 Lesser
Stag Beetle,** *Dorcus parallelopipedus*   **312 Bumble-dor,** *Geotrupes stercorarius*
313 Rose Chafer, *Cetonia aurata*   **314 Rhinoceros Beetle,** *Oryctes nasicornis*
315 Cockchafer, *Melolontha melolontha*

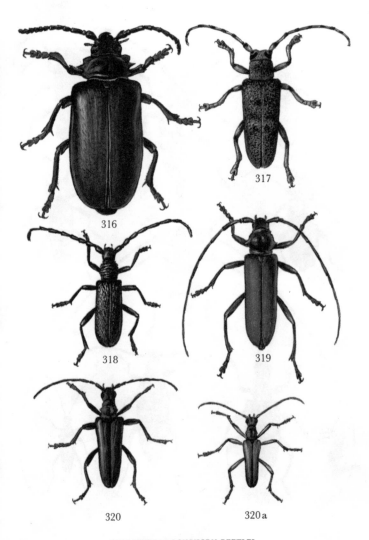

COLEOPTERA: LONGHORN BEETLES

316 *Prionus coriarius*   317 **Poplar Longicorn,** *Saperda carcharias*   318 *Cerambyx scopolii*   319 **Musk Beetle,** *Aromia moschata*   320 *Stenocorus meridianus,* female   320a male

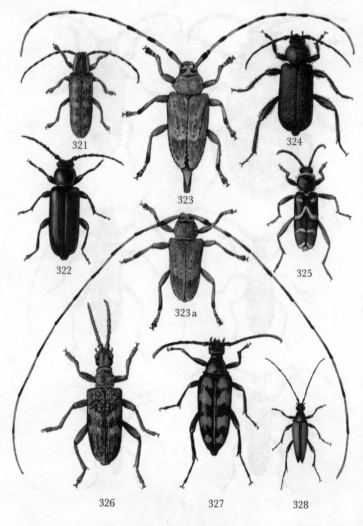

COLEOPTERA: LONGHORN BEETLES

321 *Saperda populnea* 322 *Tetropium castaneum* 323 **Timberman,** *Acantho-cinus aedilis,* female 323a male 324 *Callidium violaceum* 325 **Wasp Beetle,** *Clytus arietis* 326 *Rhagium mordax* 327 *Strangalia quadrifasciata* 328 *S. melanura*

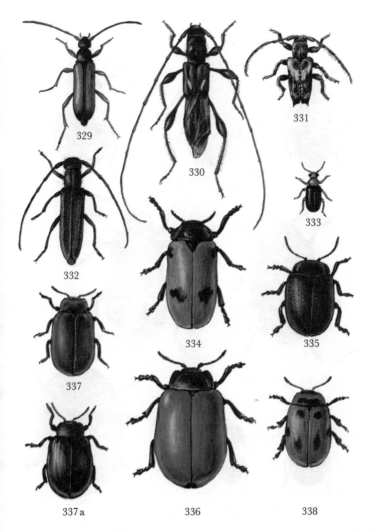

COLEOPTERA: LONGHORNS AND LEAF BEETLES
329 *Alosterna tabacicolor* 330 *Molorchus minor* 331 *Pogonocherus hispidus*
332 *Phytoecia cylindrica* 333 *Zeugophora subspinosa* 334 *Clytra quadripunctata*
335 *Chrysomela geminata* 336 *C. populi* 337 *C. aenea* 337a blue variety
338 *Phytodecta viminalis*

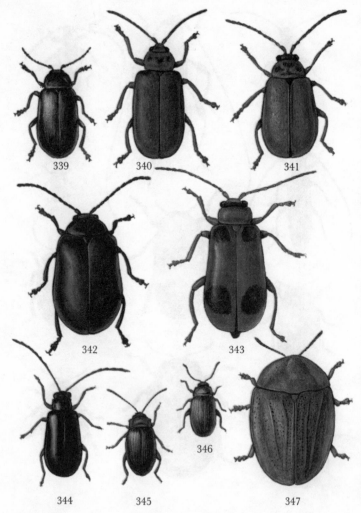

COLEOPTERA: LEAF BEETLES

339 *Phyllodecta vulgatissima* 340 *Lochmaea capreae* 341 *Galerucella lineola*
342 **Alder Leaf Beetle,** *Agelastica alni* 343 *Phyllobrotica quadrimaculata*
344 *Luperus longicornis* 345 *Derocrepis rufipes* 346 *Chalcoides fulvicornis* 347
Tortoise Beetle, *Cassida rubiginosa*

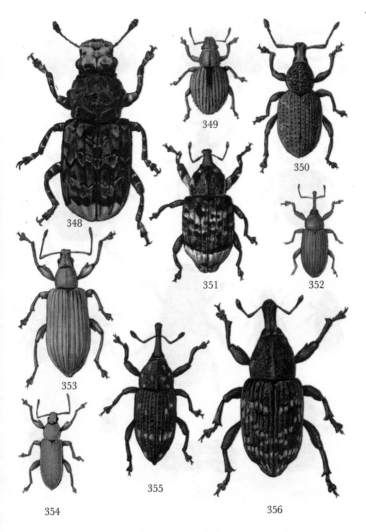

53

COLEOPTERA: WEEVILS

348 *Platyrhinus resinosus* 349 *Strophosomus melanogrammus* 350 *Otiorhynchus singularis* 351 *Cryptorhynchidius lapathi* 352 *Dorytomus tortrix* 353 *Phyllobius calcaratus* 354 *Ph. argentatus* 355 **Pine Weevil,** *Pissodes pini* 356 **Spruce Weevil,** *Hylobius abietis*

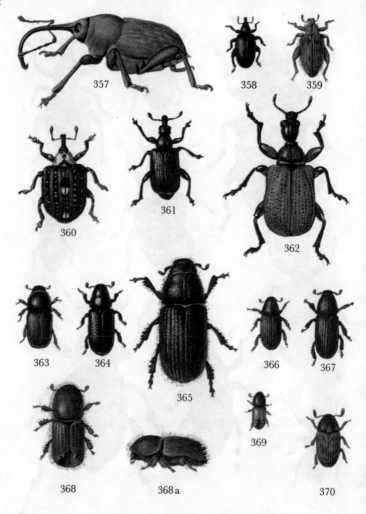

54

COLEOPTERA: WEEVILS AND BARK BEETLES

357 **Nut Weevil,** *Curculio nucum* 358 *Rhynchaenus fagi* 359 *R. quercus*
360 *Cionus scrophulariae* 361 *Rhynchites betulae* 362 *Apoderus coryli* 363
Oak Bark Beetle, *Scolytus intricatus* 364 *Blastophagus piniperda* 365 *Dendroctonus micans* 366 **Fir Bark Beetle,** *Hylurgops palliatus* 367 **Spruce Bark Beetle,** *Hylastes cunicularius* 368 **The Printer,** *Ips typographus* 369 *Pityogenes chalcographus* 370 **Ash Bark Beetle,** *Leperesinus fraxini*

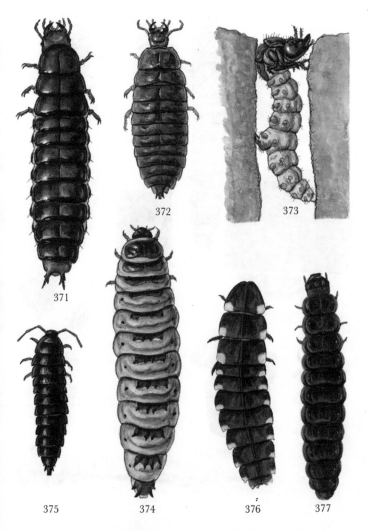

COLEOPTERA: BEETLE LARVAE

371 **Violet Ground Beetle,** *Carabus violaceus* 372 *Cychrus caraboides*, var. *rostratus* 373 **Tiger Beetle,** *Cicindela campestris* 374 **Sexton Beetle,** *Necrophorus humator* 375 **Snail Beetle,** *Phosphuga atrata* 376 **Glow-worm,** *Lampyris noctiluca* 377 **Soldier Beetle,** *Cantharis fusca*

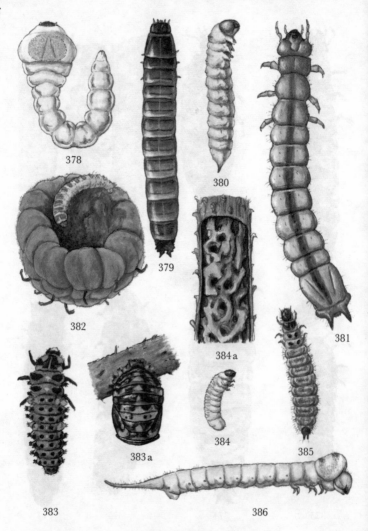

COLEOPTERA: BEETLE LARVAE

378 *Chrysobothris affinis* 379 **Wireworm,** *Denticollis linearis* 380 *Tomoxia biguttata* 381 **Cardinal Beetle,** *Pyrochroa coccinea* 382 **Raspberry Beetle,** *Byturus urbanus* 383 **Seven-spot Ladybird,** *Coccinella septempunctata* 383a pupa 384 *Ernobius mollis* 384a larval borings 385 **Ant Beetle,** *Thanasimus formicarius* 386 *Hylecoetus dermestoides*

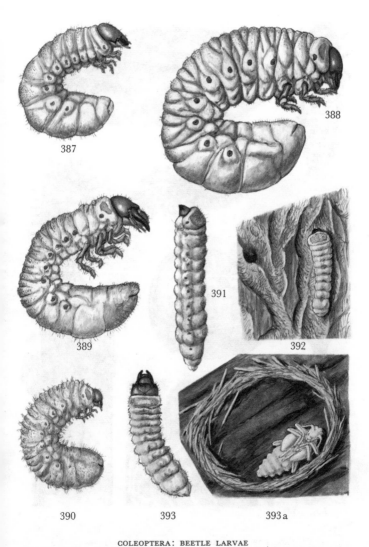

COLEOPTERA: BEETLE LARVAE
387 **Lesser Stag Beetle,** *Dorcus parallelopipedus* 388 **Rhinoceros Beetle,**
Oryctes nasicornis 389 **Cockchafer,** *Melolontha melolontha* 390 **Rose Chafer,**
Cetonia aurata 391 **Poplar Longicorn,** *Saperda carcharias* 392 *Callidium
violaceum* 393 *Rhagium mordax* 393a pupa

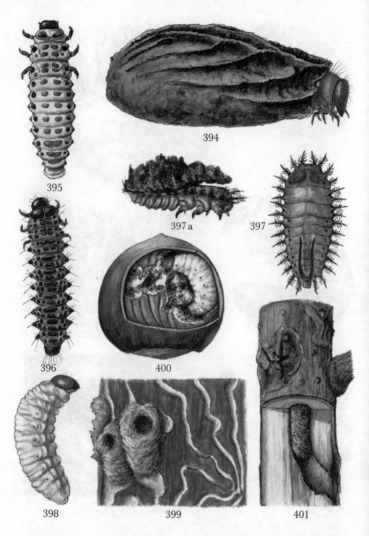

COLEOPTERA: LARVAE AND LARVAL GALLERIES
394 *Clytra quadripunctata* 395 **Poplar Leaf-beetle,** *C. populi* 396 **Alder Leaf-beetle,** *Agelastica alni* 397 **Tortoise Beetle,** *Cassida rubignosa,* seen from above without covering 397a side view with covering 398 **Spruce Weevil,** *Hylobius abietis* 399 **Pine Weevil,** *Pissodes pini* 400 **Nut Weevil,** *Curculio nucum* 401 *Cryptorhynchidius lapathi,* larval passage

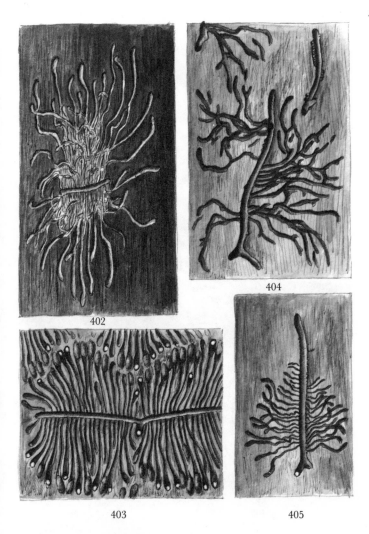

402

404

403

405

COLEOPTERA: BARK BEETLE GALLERIES
402 **Oak Bark Beetle,** *Scolytus intricatus* 403 **Ash Bark Beetle,** *Leperesinus fraxini* 404 **Fir Bark Beetle,** *Hylurgops palliatus* 405 **Spruce Bark Beetle,** *Hylastes cunicularius*

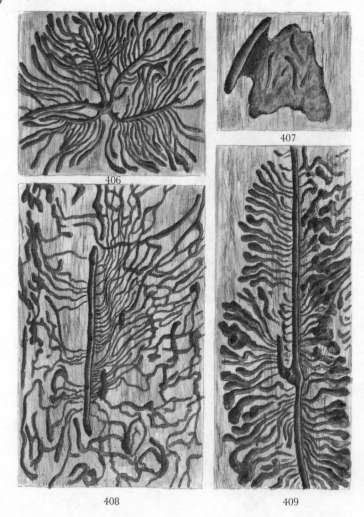

406

407

408

409

COLEOPTERA: BARK BEETLE GALLERIES
406 *Pityogenes chalcographus* 407 *Dendroctonus micans* 408 *Blastophagus piniperda* 409 **The Printer,** *Ips typographus*

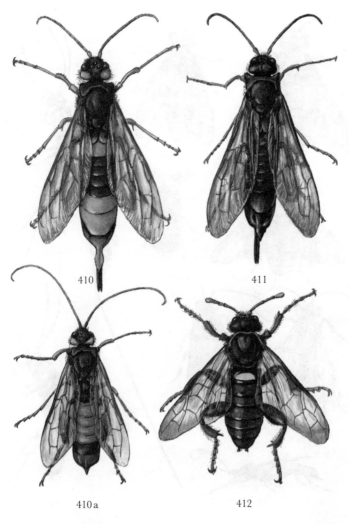

410   411

410a   412

HYMENOPTERA: WOOD WASPS AND SAWFLIES
410 **Greater Horntail,** *Sirex gigas,* female   410a male   411 **Lesser Horntail,** *S. noctilio*   412 **Birch Sawfly,** *Cimbex femorata*

62

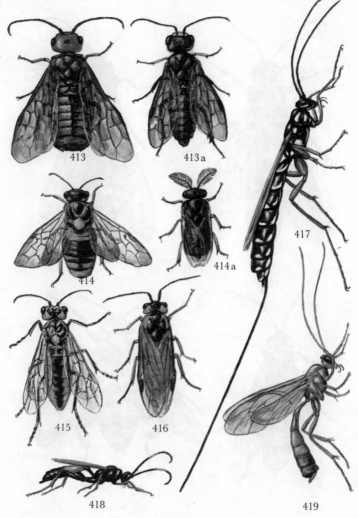

HYMENOPTERA: SAWFLIES AND ICHNEUMONS
413 **Sawfly,** *Lyda erythrocephala,* female 413a male 414 **Pine Sawfly,**
Lophyrus pini, female 414a male 415 **Sawfly,** *Rhogogaster viridis* 416
Willow Sawfly, *Pteronius salicis* 417 **Sabre Wasp,** *Rhyssa persuasoria* 418
Parasite Wasp, *Ichneumon* sp. 419 **Yellow Ophion,** *Ophion luteus*

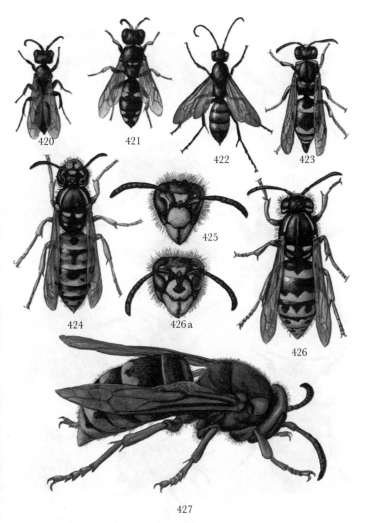

HYMENOPTERA: BURROWING AND SOCIAL WASPS
420 **Mournful Wasp,** *Pemphredon lugubris* 421 **Digger Wasp,** *Crabro vagus*
422 **Spider-hunting Wasp,** *Pompilus fuscus* 423 **Digger Wasp,** *Anchistro-*
cerus parietum 424 **Red Wasp,** *Vespa rufa* 425 **Tree Wasp,** *V. sylvestris*
426 **Common Wasp,** *V. vulgaris* 426a front view of head 427 **Hornet,**
V. crabro

64

HYMENOPTERA: BEES

428 **Large Red-tailed Bumble-bee,** *Bombus lapidarius,* female 428a worker bee 428b male 429 **Early Bumble-bee,** *B. pratorum,* female 429a male 430 **Buff-tailed Bumble-bee,** *B. terrestris,* female 430a male 431 **Common Carder-bee,** *B. agrorum,* female 431a male 432 **Vestal Cuckoo-bee,** *Psithyrus vestalis,* female 432a male 433 **Blue Osmia,** *Osmia caerulescens*

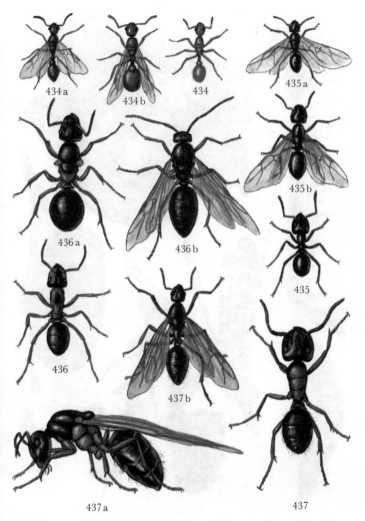

HYMENOPTERA: ANTS

434 **Red Ant,** *Myrmica laevinodis,* worker ant   434a male   434b female
435 **Orange Ant,** *Lasius fuliginosus,* worker ant   435a male   435b female
436 **Wood Ant,** *Formica rufa,* worker ant   436a female without wings,
436b male   437 **Hercules Ant,** *Camponotus herculeanus,* worker ant   437a
female, 437b male

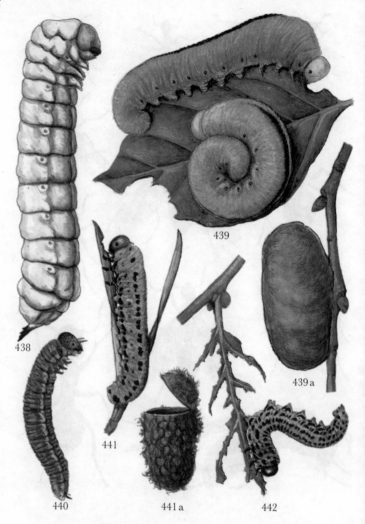

HYMENOPTERA: LARVAE OF WOOD WASPS AND SAWFLIES
438 **Greater Horntail,** *Sirex gigas* 439 **Birch Sawfly,** *Cimbex femorata*
440 **Sawfly,** *Lyda erythrocephala* 441 **Pine Sawfly,** *Lophyrus pini* 441a cocoon
442 **Willow Sawfly,** *Pteronus salicis*

67

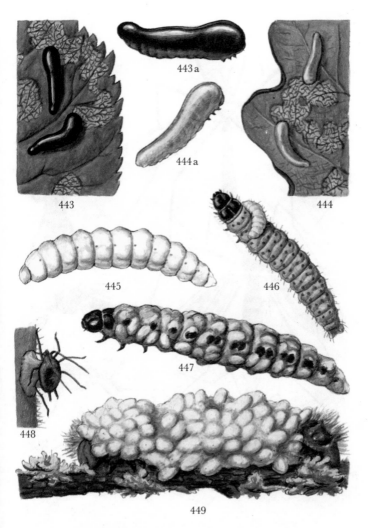

HYMENOPTERA: LARVAE OF SAWFLIES, ICHNEUMONS, ETC.

443 **Pear Sawfly,** *Eriocampoides limacina* 443a greatly enlarged, side view
444 **Oak Sawfly,** *E. annulipes* 444a greatly enlarged, side view 445 **Sabre
Wasp,** *Rhyssa persuasoria* 446 Larva of wasp parasitic upon weevil larva
447 Moth caterpillar containing pupae of parasitic wasp. 448 Aphid with
cocoon of braconid, *Praon* sp. 449 Parasitized caterpillar on spruce with
pupae of a braconid, *Microgaster* sp.

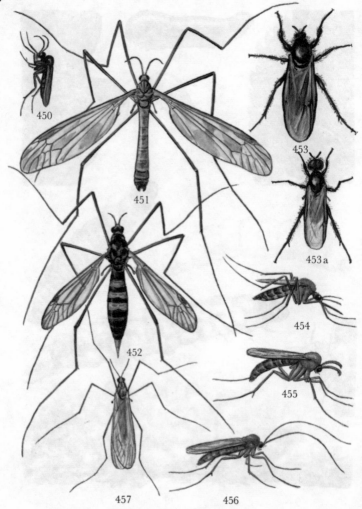

DIPTERA: GNAT-LIKE FLIES

450 **Fungus Gnat,** *Sciara* sp. 451 **Crane-fly,** *Tipula nebuculosa* 452 **Crane-fly,** *Nephrotoma crocata* 453 **St. Mark's Fly,** *Bibio marci,* female 453a male
454 **Biting Gnat,** *Aedes* sp. 455 **Fungus Gnat,** *Mycetophilidae* sp. 456 **Long-horned Fungus Gnat,** *Macrocera* sp. 457 **Winter Gnat,** *Trichocera hiemalis*

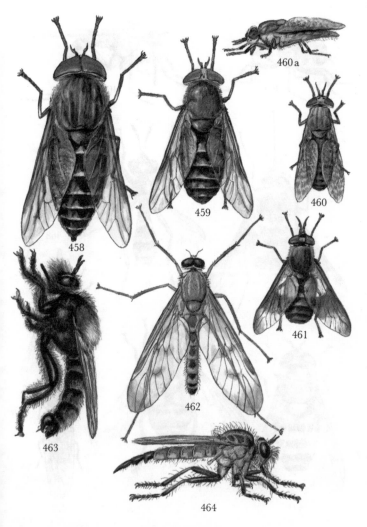

DIPTERA: FLIES

458 **Horse-fly,** *Tabanus bovinus* 459 **Cleg,** *Hybomitra collina* 460 **Cleg,** *Haematopota pluvialis* 460a side view 461 **Cleg,** *Chrysops pictus* 462 **Snipe-fly,** *Rhagio scolopacea* 463 **Robber-fly,** *Laphria ephippium* 464 **Robber-fly,** *Neoitamus cyanurus*

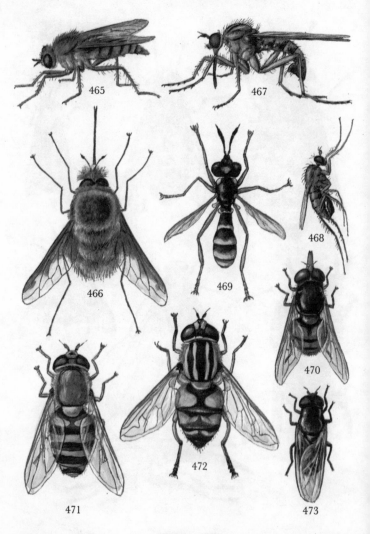

DIPTERA: FLIES

465 **Stiletto-fly,** *Thereva nobilitata*   466 **Greater Bee-fly,** *Bombylius major*
467 **Empid,** *Empis tesselata*   468 **Long-headed Fly,** *Dolichopus claviger*
469 **Wasp Fly,** *Conops quadrifasciata*   470 **Hover-fly,** *Rhingia campestris*
471 **Common Hover-fly,** *Syrphus ribesii*   472 **Hover-fly,** *Helophilus pendulus*
473 **Hover-fly,** *Chilosia albitarsis*

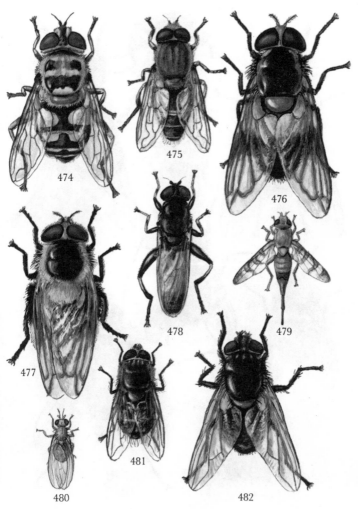

DIPTERA: FLIES

474 **Hover-fly,** *Myiatropa florea*   475 **Drone-fly,** *Eristalis* sp.   476 **Hover-fly**
Volucella pellucens   477 **Hover-fly,** *V. bombylans*   478 **Hover-fly,** *Xylota*
segnis   479 **Fruit-fly,** *Ceriocera ceratocera*   480 **Marsh-fly,** *Lyciella rorida*
481 **Flesh-fly,** *Pollenia rudis*   482 **Muscid,** *Mesembrina meridiana*

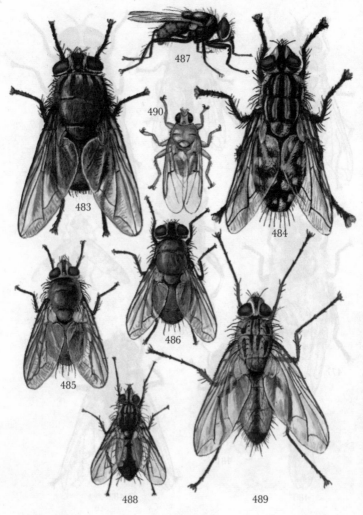

DIPTERA: FLIES

483 **Blow-fly,** *Calliphora vomitoria* 484 **Flesh-fly,** *Sarcophaga carnaria* 485
Parasite-fly, *Gymnochaeta viridis* 486 **Green-bottle,** *Lucilia* sp. 487
Muscid, *Hydrotaea irritans* 488 **Parasite-fly,** *Eriothrix rufomaculatus* 489
Parasite-fly, *Dexia rustica* 490 **Bird Louse Fly,** *Ornithomyia avicularia*

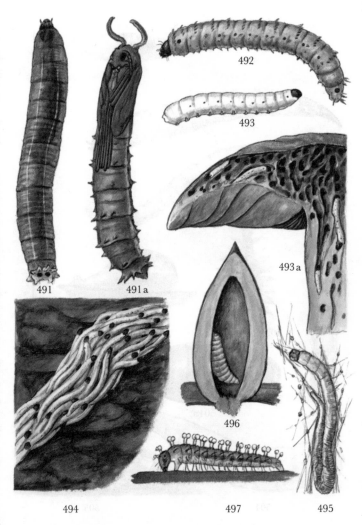

DIPTERA: LARVAE OF GNAT-LIKE FLIES

491 **Crane-fly,** *Tipula* sp., larva 491a pupa 492 **St. Mark's Fly,** *Bibio marci* 493 **Fungus Gnat,** *Mycetophilidae* sp. 493a larvae in fungus 494 **Army Worm,** *Sciara* sp. 495 **Long-horned Fungus Gnat,** *Macrocera* sp. 496 **Gall Gnat,** *Mikiola fagi,* larva in hollow gall 497 **Biting Midge,** *Forcipomyia* sp.

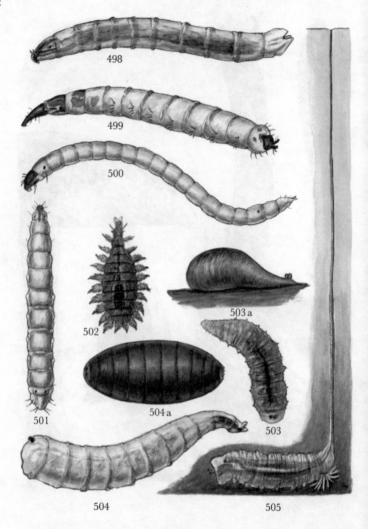

DIPTERA: MISCELLANEOUS FLY LARVAE

498 **Snipe-fly,** *Rhagio* sp.   499 **Snipe-fly,** *Xylophagus* sp.   500 **Stiletto-fly,**
Thereva sp.   501 **Robber-fly,** *Neoitamus* sp.   502 **Muscid,** *Fannia* sp.
503 **Hover-fly,** *Syrphus* sp.   504 **Blow-fly,** *Calliphora* sp., larva   504a pupa
505 Rat-tailed maggot of **Drone-fly,** *Eristalis* sp.

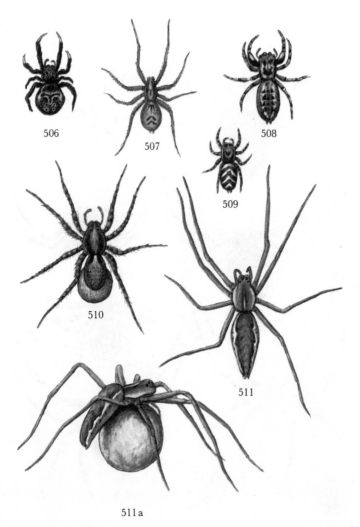

ARANEAE: SPIDERS

506 *Hyptiotes paradoxus* 507 *Anyphaena accentuata* 508 **Jumping Spider,**
Marpissa muscosa 509 **Zebra Spider,** *Salticus scenicus* 510 **Wolf Spider,**
Lycosa amentata 511 **Hunting Spider,** *Pisaura mirabilis,* male 511a female
with egg sac

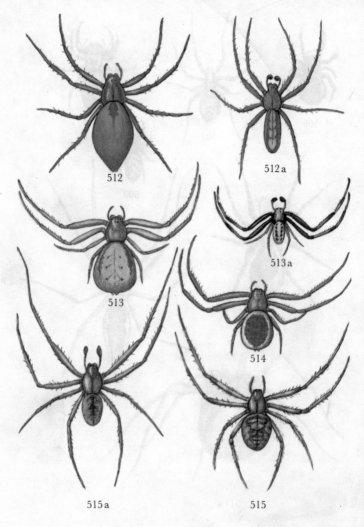

ARANEAE: SPIDERS
512 *Micrommata virescens*, female   512a male   513 **Crab Spider,** *Misumena vatia*, female   513a male   514 *Diaea dorsata*, female   515 *Meta segmentata*, female   515a male

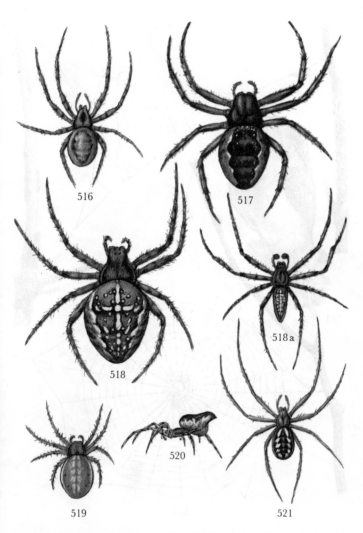

ARANEAE: SPIDERS

516 *Zygiella atrica* 517 *Araneus umbraticus* 518 **Cross Spider,** *A. diadematus,*
female 518a male 519 *A. cucurbitinus* 520 *Cyclosa conica* 521 *Linyphia
triangularis*

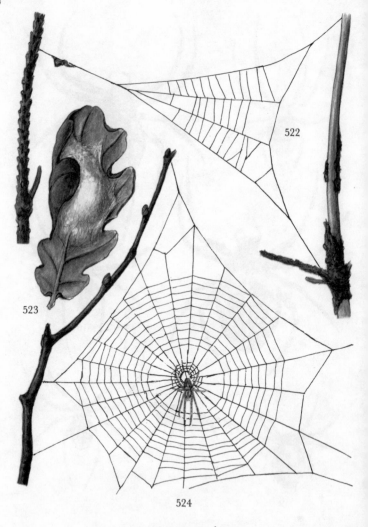

ARANEAE: SPIDERS' WEBS
522 *Hyptiotes paradoxus,* snare 523 *Anyphaena accentuata,* egg cocoon 524
Meta segmentata, snare

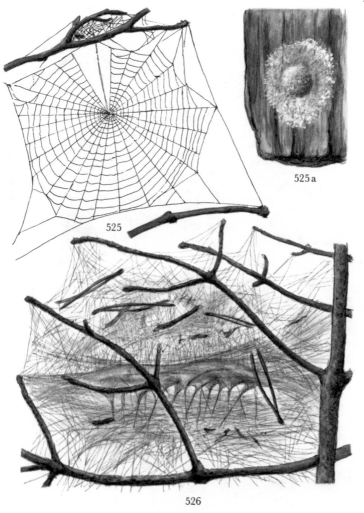

525

525a

526

ARANEAE: SPIDERS' WEBS
525 *Zygriella atrica*, snare   525a egg sac   526 *Linyphia triangularis*, snare

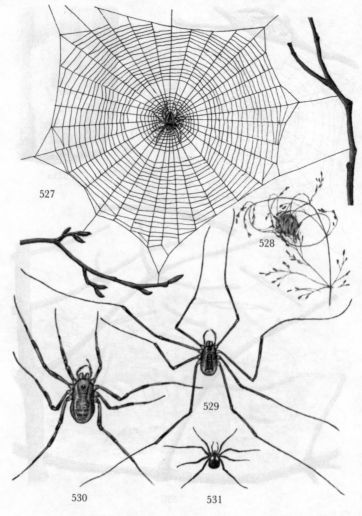

ARANEAE AND OPILIONES: SPIDERS' WEBS AND HARVEST-SPIDERS
527 **Cross Spider,** *Araneus diadematus* 528 *A. cucurbitinus,* egg cocoon
529 **Harvestman,** *Liobunum rotundum* 530 **Harvestman,** *Lacinius ephippiatus*
531 **Harvestman,** *Nemastoma lugubre*

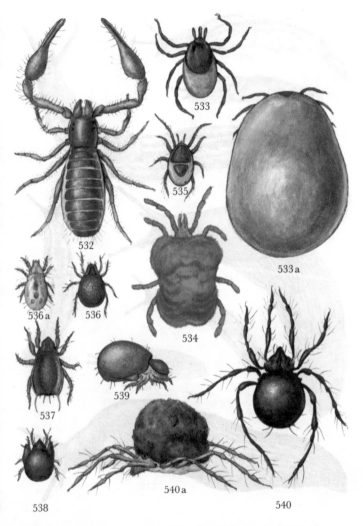

CHELONETHI AND ACARI: PSEUDO-SCORPIONS, MITES AND TICKS

532 **Pseudo-scorpion,** *Neobisium muscorum*   533 **Sheep Tick,** *Ixodes ricinus,*
before and   533a after feeding   534   **Red Earth Mite,** *Trombidium holoceri-*
ceum   535 **Mite,** *Parasitus* sp.   536 **Mite,** *Banksia tegeocrana*   536a larva
537 **Mite,** *Camisia palustris*   538 **Mite,** *Galumna climata*   539 **Mite,** *Phthira-*
carus sp.   540 **Oribatid Mite,** *Oribata geniculata*   540a larva with covering

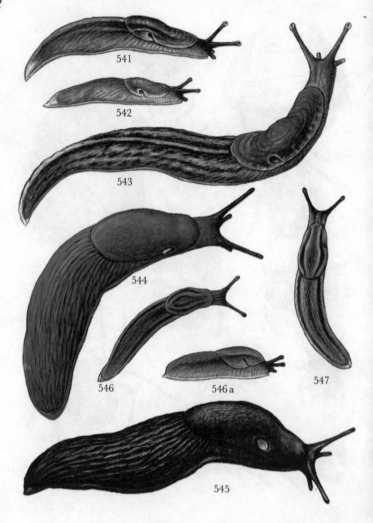

MOLLUSCA: SLUGS

541 *Limax marginatus* 542 *L. tenellus* 543 *L. maximus* 544 **Red Slug,** *Arion rufus* 545 **Black Slug,** *A. ater* 546 **Brown Slug,** *A. subfuscus* 546a side view 547 *A. circumscriptus*

83

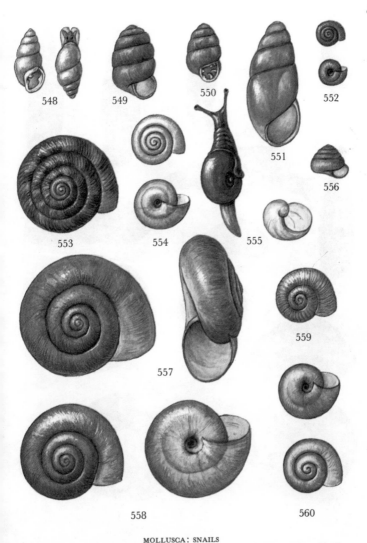

MOLLUSCA: SNAILS

548 *Carychium tridentatum* 549 *Columella edentula* 550 **Chrysalis Snail,**
Vertigo sp. 551 *Cochlicopa lubrica* 552 *Punctum pygmaeum* 553 *Discus
rotundatus* 554 **Crystal Snail,** *Vitrea crystallina* 555 **Glass Snail,** *Vitrina
pellucida* 556 *Euconulus fulvus* 557 *Retinella nitidula* 558 *Oxychilus allvarium*
559 *Perpolita hammonis* 560 *Retinella pura*

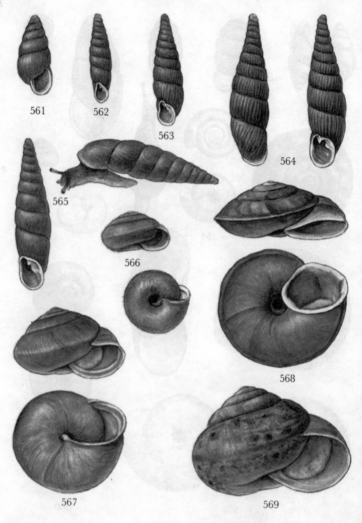

MOLLUSCA : SNAILS

561 *Ena obscura*   562 **Little Coiled Snail,** *Clausilia bidentata*   563 **Common
Coiled Snail,** *C. pumila*   564 **Large Coiled Snail,** *Iphigena ventricosa*   565
Smooth Coiled Snail, *Cochlodina laminata*   566 **Hairy Garden Snail,**
Trichia hispida   567 **Hazel Snail,** *Perforatella incarnata*   568 *Helicogonia
lapicida*   569 **Bush Snail,** *Eulota fruticum*

MOLLUSCA: SNAILS
570 *Arianta arbustorum* 571 **White-lipped Hedge-snail,** *Cepaea hortensis*
571a Underside of shell, showing lip 572 **Black-lipped Hedge-snail,** *C. nemoralis* 572a underside of shell, showing lip 572b and 527c colour variants 573 **Roman Snail,** *Helix pomatia* 573a Epiphragm

FERN AND SPRUCE

574 Intwined point on fern frond (**muscid,** *Chirosia parvicornis*)   575 Gall on the branch of spruce (**aphid,** *Gilletteella cooleyi*)   576 Small pineapple gall on spruce (**aphid,** *Cnaphalodes* sp.)   576a old gall   577 Pseudocone gall on spruce (**aphid,** *Adelges abietis*)   578 Deformed needles (**aphid,** *Mindarus abietinus*)   579 Brown needles (**moth,** *Eucosma tedella*)

SPRUCE AND PINE
580 Bent top shoots and attacked cones (**moth,** *Dioryctria abietella*) 580a
Extruded faeces 581 Stripped bark (**weevil,** *Hylobius abietis*) 582
Whitened needles (**gall-gnat,** *Thecodiplosis brachyntera*) 582a needles enlarged
583 Matted needles (**sawfly,** *Lyda erythrocephala*)

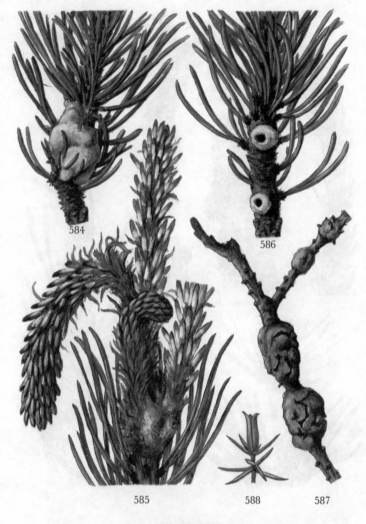

PINE AND JUNIPER
584 Resin 'galls' (**moth,** *Evertia resinella*)   585 Drooping top shoot (**moth,** *Evertia buoliana*)   586 Red needles on pine (**beetle,** *Blastophagus piniperda*)
587 Green gall on pine (**mite,** *Eriophyes pini*)   588 Juniper 'berries' (**gall-gnat,** *Oligotrophus juniperinus*)

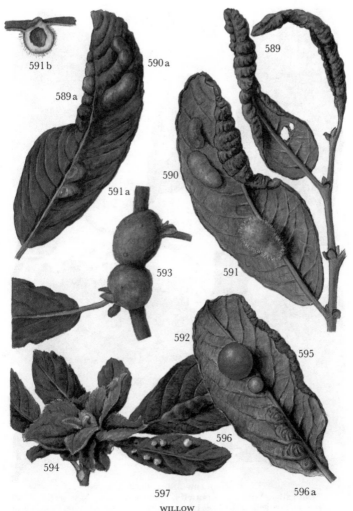

WILLOW

589 Both edges of leaf curled (**sawfly,** *Pontania leucaspis*) 589a single edge curled 590 Elongated leaf gall on upper surface (**sawfly,** *P. vesicator*) 590a under surface 591 Hairy leaf gall on under surface (**sawfly,** *P. pendunculi*) 591a upper surface 591b section 592 Spherical leaf gall (**sawfly,** *P. viminalis*) 593 Globose gall on branch (**gall-gnat,** *Rhabdophaga salicis*) 594 Camellia gall (**gall-gnat,** *Rh. rosaria*) 595 Rolled leaf margin (**gall-gnat,** *Dasyneura* sp.) 596 Pouch gall on lamina (**gall-gnat,** *Iteomyia capreae*) 596a opening on under surface 597 Small hairy gall on lamina (**gall-mite,** *Eriophyes tetanothrix*)

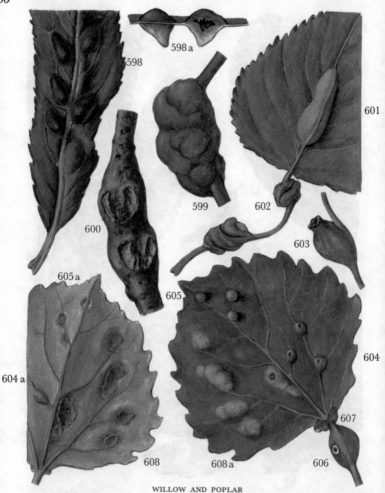

WILLOW AND POPLAR

598 Bean gall on willow leaf (**sawfly,** *Pontania capreae*) 598a section 599 Gherkin gall on willow stem (**gall-wasp,** *Euura pentandrae*) 600 Branch gall on poplar (**beetle,** *Saperda populnea*) 601 Midrib gall on poplar (**aphid,** *Pemphigus filaginis*) 602 Spiral gall on poplar petiole (**aphid,** *P. spirothecae*) 603 Purse gall on poplar peticle (**aphid,** *P. bursarius*) 604 Vein gall on upper surface on poplar leaf (**gall-gnat,** *Harmandia cavernosa*) 604a underside 605 Vein gall on upper surface of poplar leaf (**gall-gnat,** *H. globuli*) 605a underside 606 Gall on poplar petiole (**gall-gnat,** *Syndiplosis petioli*) 607 Swollen leaf glands on poplar (**mite,** *Eriophyes diversipunctatus*) 608 Hairy patches on under side of poplar leaf (**mite,** *E. varius*) 608a upper surface

ALDER, HAZEL AND BIRCH

609 Small leaf gall on alder (**mite,** *Eriophyes laevis*)   610 Hairy spots on underside of alder leaf (**mite,** *E. brevitarsus*)   611 Roll-gall on hazel leaf (**weevil,**
Apoderus coryli)   612 Roll-gall on birch leaf (**weevil,** *Rhynchites betulae*)
613 Leaf-miner on birch (**moth,** *Eriocrania sparmanella*   614 Leaf-miner on
birch (**moth,** *Lyonetia clerckella*)   615 Leaf-miner on birch (**fly,** *Agromyza
alni-betulae*)

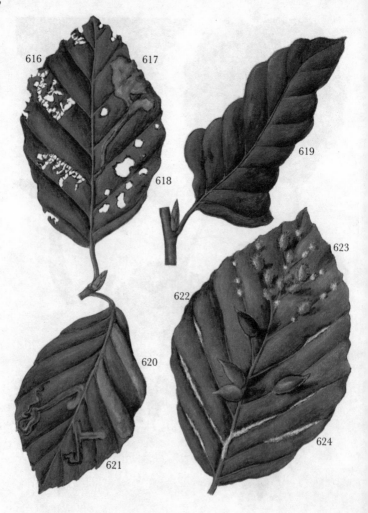

BEECH

616 Small holes in leaf (**weevil,** *Rhynchaenus fagi*) 617 Leaf-miner (*R. fagi*)
618 Large holes in leaf (**weevil,** *Phyllobius argentatus*) 619 Roll-gall on leaf
(**aphid,** *Phyllaphis fagi*) 620 Leaf-miner (**moth,** *Lithocolletis faginella*)
621 Leaf-miner (**moth,** *Nepticula basalella*) 622 Pointed gall on leaf (**gall-gnat,** *Mikiola fagi*) 623 Hairy gall on leaf (**gall-gnat,** *Hartigiola annulipes*)
Vein gall on leaf (**mite,** *Eriophyes nervisequus*)

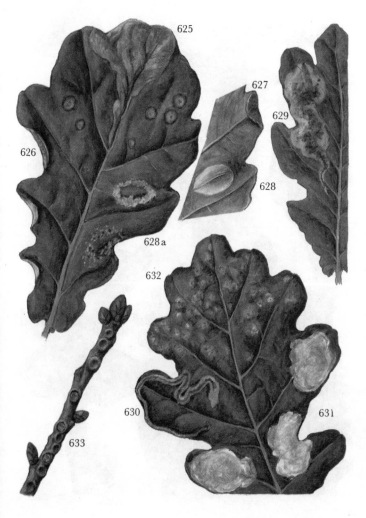

OAK

625 Leaf-miner (**weevil,** *Rhynchaemus quercus*) 626 Folded leaf margin (**gall-gnat,** *Macrodiplosis volvens*) 627 Folded leaf margin (**gall-gnat,** *M. dryobia*)
628 Wrinkled mine on under surface of leaf (**moth,** *Lithocolletis* sp.) 628a
upper surface 629 Leaf-miner (**sawfly** *Profenusa pygmaea*) 630 Leaf-miner
(**moth,** *Nepticula* sp.) 631 Leaf-miners (**moth,** *Tishceria* sp.) 632 Yellow
spots on lamina (**aphid,** *Phylloxera quercus*) 633 Pit gall on stem (**coccid,**
Asterolecanium variolosum)

94

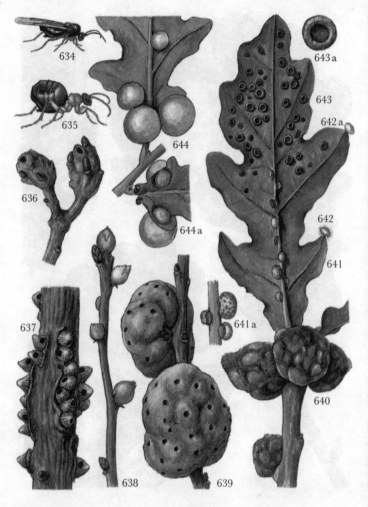

OAK

634 **Gall wasp,** *Neuroterus* sp. winged female 635 **Gall-wasp,** *Biorhiza pallida,* wingless female 636 Gall on shoot apex (**gall-wasp,** *Andricus inflator*) 637 Bark gall (**gall-wasp,** *A. testaceipes*) 638 Bud gall (**gall-wasp,** *Trigonaspis megaptera*) 639 Oak apple (**gall-wasp,** *Biorhiza pallida*) 640 Artichoke gall (**gall-wasp,** *Andricus fecundatrix*) 641 Oyster gall (**gall-wasp,** *A. ostreus*) 641a enlarged 642 Leaf edge hairy gall (**gall-wasp,** *Neuroterus albipes*) 642a later stage (smooth) 643 Silk-button gall (**gall-wasp,** *N. numismatis*) 643a enlarged 644 Currant gall on under surface of leaf (**gall-wasp,** *N. quercus-baccarum*) 644a upper surface

95

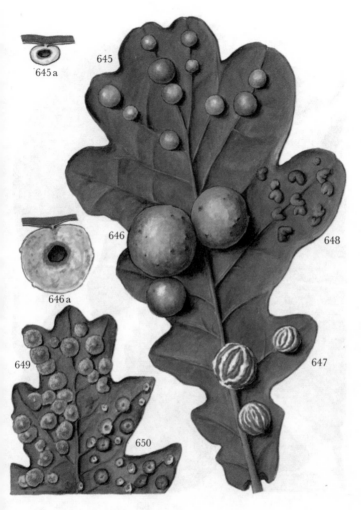

OAK

645 Red pea-gall (**gall-wasp,** *Diplolepis divisa*)   645a section   646 Cherry gall (**gall-wasp,** *D. quercus-folii*)   646a section   647 Striped gall (**gall-wasp,** *D. longiventris*)   648 Kidney gall (**gall-wasp,** *Trigonaspis megaptera*)   649 Common spangle-gall (**gall-wasp,** *Neuroterus quercus-baccarum*)   650 Smooth spangle-gall (**gall-wasp,** *N. albipes*)

ELM, NETTLE, RASPBERRY

651 Rolled margin of elm leaf (**aphid,** *Eriosoma ulmi*) 652 Vein gall on elm leaf (**aphid,** *Tetraneura ulmi*) 652a enlarged 653 Leaf-miner on elm (**moth,** *Lithocolletis* sp.) 654 Leaf-miner on elm (**moth,** *Nepticula* sp.) 655 Brown spot on elm leaf (**mite,** *Eriophyes filiformis*) 656 Gall on stinging nettle (**gall-gnat,** *Dasyneura urticae*) 657 Stem gall on raspberry (**gall-gnat,** *Lasioptera rubi*) 658 Stem gall on dewberry (**gall-wasp,** *Diastrophus rubi*)

ROSE, ETC.
659 Bedeguar gall on rose (**gall-wasp** *Diplolepis rosae*)   660 Spiked pea-gall
on rose (**gall-wasp,** *D. rosarum*)   661 Smooth pea-gall on rose (**gall-wasp,** *D. eglanteriae*)   662 Oblong leaf gall on rose (**gall-wasp,** *D. spinosissimae*)   633
Rolled leaflet margin on rose (**sawfly,** *Blennocampa pusilla*)   664 Defoliator on
rose (**sawfly,** *Cladius pectinicornia*)   665 Leaf gall on meadowsweet (**gall-gnat,**
Perrisia ulmariae)

98

ROSACEOUS TREES, ETC.

666 Curled leaves at shoot apex of cherry (**aphid,** *Myzus cerasi*) 667 Pustulate galls on blackthorn (**mite,** *Eriophyes similis*) 668 Rolled leaf margins on hawthorn (**mite,** *E. goniothorax*) 669 Leaf rosette on hawthorn (**gall-gnat,** *Perrisia cratoegi*) 670 Rolled leaf margin on crab-apple (unidentified **moth**) 671 Brown spots on crab-apple (**mite,** *Eriophyes piri*) 672 Leaf-miner on balsam (**fly,** *Liriomyza impatientis*) 673 Leaf-miner on labiate (**fly,** *Phytobia labiatarum*)

LIME AND SYCAMORE

674 Globose galls on lime (**mite,** *Eriophyes tetratrichus*) 675 Brown spots on lime (**mite,** *E. liosoma*) 676 Nail galls on lime (**mite,** *E. tiliae*) 677 Leaf-vein galls on sycamore (**mite,** *E. macrochelus*) 678 Hairy pouch galls on sycamore (**mite,** *E. megalonyx*) 679 Red pustulate galls on sycamore (**mite,** *E. macrorhynchus*) 680 Rolled leaf margins on sycamore (**weevil,** *Acleris sponsana*)

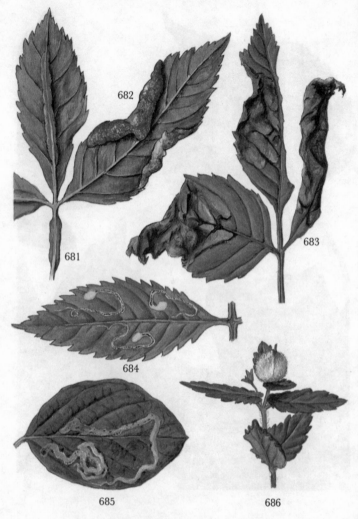

ASH, HONEYSUCKLE AND SPEEDWELL

681 Midrib gall on ash (**gall-gnat,** *Perrisia fraxini*) 682 Rolled leaf margin on
ash (psyllid *Psyllopsis fraxini*) 683 Leaf-miner on ash (**moth,** *Gracilaria
syringella*) 684 Leaf-miner on ash (**moth,** *Praus curticellus*) 685 Leaf-miner
on honeysuckle (**fly,** *Phytagromyza xylostei*) 686 Shoot tip gall on speedwell
(**gall-gnat,** *Jaapiella veronicae*)

WOODS

Woods are of several different types, each of which has its own characteristic fauna in the sense of harbouring a wide range of organisms, especially insects, which are so closely connected with particular plant species that they cannot live where the host plants are absent. It is easy to understand, therefore, that the greater the variety of trees, bushes and herbs growing in the wood, the greater the variety of animals to be found there. Large areas of forest, consisting solely of beech or fir, contain only a limited number of species, even though individuals of a particular species may be present in very large numbers. Furthermore, there are also organisms characteristic of a wood which are not associated with any particular plant, but are able to survive only under the particular biological conditions prevailing in a wood. Because of the trees, and the shade and shelter they provide, woods differ in important respects from open country. The light in them is more subdued – the leaf canopy of spruce and beechwoods is often so thick as to admit only a small percentage of free daylight; the air is damper; the daily temperature range is less; and the soil is generally different in quality. Types of woods are summarised below. They are divided into two main groups; coniferous softwoods and broad-leaved deciduous hardwoods.

Coniferous woods are characterised by the fact that certain groups of organisms, such as worms and snails, are either rare or absent altogether, whereas other species, like ants and certain spiders, are more conspicuous than in deciduous woods.

A spruce wood is poor in species of animals and plants owing to the fact that the trees usually stand so close together and exclude so much light that there is a scarcity of undergrowth on the forest floor. As a rule, therefore, the only animals colonising spruce forests are those associated with the spruce trees themselves or which live in the soil. Pine woods are generally richer in species because the pine trees seldom grow so close together or exclude so much light as spruce, so that bushes and various herbaceous plants tend to grow thickly beneath them. Consequently it is possible to find in pine forests not only species connected with the pine trees themselves but also those whose life is conditioned by the undergrowth. In larch woods, which are seldom extensive, occurring mainly as small plantations, much the same applies as to pine woods. Mountain pine plantations are a particular type of pine forest, occurring mainly on very poor soil. In such plantations the fauna is low in species and inconspicuous.

One form of hardwood forest is a beech wood. Beech forests can be divided into a series of types according to the kind of soil and the condition of the wood. These types are characterised by the predominating plants among the flora growing on the forest floor, and are not, of course, sharply delimited, but merge gradually one into another. In general the poorest fauna is to be found in the beech woods on the poorest soil.

Characteristic plants among the ground flora include the woodsorrel, chickweed, wintergreen and various grasses. The last mentioned are as a rule predominant where there is a strong flow of air through the wood. In this type of beechwood, the only organisms to be seen, as a rule, are those connected with the beech itself or living on the forest floor. Where the wood is growing on better soil there may be anemones in addition to the sorrel, and the ground becomes covered with a layer of more or less rotted dead leaves which often harbour a rich animal life. It is in this type of wood that earthworms and snails are common. Such a fauna, however, is better developed in the more open beech woods, where the herb layer consists largely of anemones and woodruff, sometimes with an admixture of enchanter's nightshade (*Circaea*), dog's mercury, etc. In these, as a rule, the ground is deeply covered with mould, and the fallen leaves are soon decomposed by organisms living upon them. Snails, too, are common, both on the forest floor and crawling over the trunks.

Oak woods, in contrast to beech woods, often contain a rich scrub layer of smaller trees and bushes and they also have a greater range of herbaceous plants on the forest floor. This is due to the relatively open tops of the oak trees, which allow more light to reach the interior than the dense canopy of the beech. As there are many more organisms directly connected with oak than with beech, and as the undergrowth is more varied, the fauna in oak woods is the richer of the two.

The richest fauna of all exists, as one might expect, in mixed deciduous forests containing oak, ash, sycamore and elm as important trees, occasionally with beech and lime, and in damper situations, willow and alder. When a mixed wood grows on relatively good soil, there will usually be a rich and varied undergrowth; most of the fauna connected with deciduous forests is to be found in woods of this type.

DESCRIPTIONS AND NOTES

OLIGOCHAETA - Earthworms

1 Pot Worm (*Enchytrae*)
Mesenchytraeus setosus
1-2 cm long. Often abundant under fallen leaves and in the soil where there may be many thousands per square metre. There are several species which closely resemble one another.

2 Marsh Worm. *Lumbricus rubellus*
10-12 cm long. As in other species of *Lumbricus*, the tail may be expanded into a flat 'trowel'. Most frequent in soil containing rich mould, although it also occurs in poorer ground. The larger species of earthworms live partly on organic matter in the soil and partly on dead leaves which they pull down a short way into their holes (2a). Their droppings (worm casts), may often be seen on the surface (2b).

3 *L. castaneus*
7 cm long. Easily recognized by the 'saddle' (*clitellum*) being almost in the middle. Frequent underneath leaf-mould lying on rich soil. (English Editor's note: Tail is *not* normally flattened as in the illustration.)

4 *Allolobophora turgida*
10-15 cm long. Distinguished from *Lumbricus* (2) by the tail not being flattened and by the colour not being irridescent. This is perhaps the commonest earthworm in woods, fields, and gardens alike, particularly in situations where the soil contains an abundance of well-rotted leaf-mould.

5 *Dendrobaena octaedra*
3-4 cm long. One of the few earthworms found on damp ground in coniferous woods. Often encountered in moss or beneath stones.

6 *D. arborea*
4-6 cm long; often very slender. Common in decaying tree stumps and may sometimes be found high up on trees inside rotten knots. Frequently seen in damp weather creeping over the surface of dead wood.

7 *Octolasium cyaneum*
10-18 cm long. Recognised by its delicate pink and lilac colouring. Common in woods beneath slightly damp leaf-litter.

CRUSTACEA: Woodlice

8 Woodlouse, *Ligidium hypnorum*
10 mm long. Being crustaceans, wood-lice are animals of damp situations. They live mainly on decaying plants. *Ligidium* is to be found on any damp ground in hardwood forests and moves in a manner not usual among woodlice

9 Woodlouse, *Trichoniscus pusillus*
2-5 mm long. Very common on damp ground within hardwood forests. May easily be confused with the young of other woodlice.

10 Woodlouse, *Philoscia muscorum*
10 mm long. Apart from the forest floor, it may also be seen on vegetation above ground. Particularly common on raspberry plants.

11 Woodlouse, *Oniscus asellus*
Up to 17 mm long. Very common among stones and plant debris in woods and gardens and in similar situations where the soil is damp.

12 Woodlouse, *Porcellio scaber*
Up to 17 mm long. Very common in woods and other sheltered places, especially under stones and bark. It may also invade houses and sometimes feeds high up on trees.

13 Pill Woodlouse, *Armadillidium cinereum*
Up to 18 mm long. Particularly com-mon on chalky ground. Found near houses and on the fringes of woods more often than in the woods them-selves. When disturbed rolls itself up into a ball (13a).

DIPLOPODA : Millepedes

14 Pill Millepede, *Glomeris marginata*
Up to 20 mm long. Common in leaf litter on the floor of deciduous forests, especially beechwood. When dis-turbed, rolls itself into a ball (14a) like the Pill Woodlouse, from which it may be distinguished by the undivided tail

15 Flat-back Millepede, *Polydesmus complanatus*
Up to 20 mm long. Common in hard-wood forests with closely related, smaller species. The male has 30 and the female 31 pairs of legs.

16 Woodland Snake Millepede, *Cylindroiulus sylvarum*
Up to 30 mm long. Very common on the floor both of deciduous and conif-erous woodlands. The number of segments, and therefore of limbs, is variable, but there are generally about 100 pairs of legs. Although millepedes live mainly on decaying vegetation, they also eat into such soft fruits as strawberries.

17 Dwarf Millepede, *Scutigerella immaculata*
Up to 8 mm long. Not a true millepede as it has only twelve pairs of legs. Very common under stones, the bark of stumps, or fallen branches on the forest floor.

CHILOPODA : Centipedes

18 Lithobiomorph Centipede, *Lithobius forficatus*
Up to about 30 mm long. Very com-mon under stones or bark both in woods and elsewhere. Each of the fore-most pair of legs is modified to form a dagger-shaped poison-claw whereby the prey, which consists of insect larvae or the like, is seized and killed. Does not become mature until three years old.

19 Geophilomorph Centipede,
Geophilus sp.
Up to 45 mm long. Several species resembling one another are found under bark, or stones in woods. Some kinds are luminous. The female lays her eggs in a cluster and winds herself round them, remaining there until the eggs are hatched.

COLLEMBOLA : Springtails

20 Springtail, *Hypogastrura armata*
About 1 mm long. A springtail receives its popular name from the fork-shaped apparatus whereby it jumps, which extends from the point of the back along the underside of the abdomen. The animal flings itself upwards and forwards by thrusting the fork downwards.

21 Springtail, *Neanura muscorum*
2 mm long. Like *Hypogastrura* it is abundant in woodland soil where there may be as many as 7000 per square metre.

22 Springtail, *Onychiurus armatus*
2.5-3 mm long. The white colour and the absence of eyes show this to be an animal living in the dark. Found especially in damp ground under deciduous refuse.

23 Springtail, *Isotoma viridis*
3 mm long. Common in damp ground under deciduous trees. By reason of their large numbers they play an important role in decomposing leaf litter.

24 Springtail, *Orchesella flavescens*
5-6 mm long. Very variable in marking. Common on and beneath rotten branches on the forest floor.

25 Springtail, *Tomocerus plumbeus*
5-6 mm long. Abundant in the forest floor. Adults are first encountered during July. They die off in the autumn and over-winter as eggs.

26 Springtail, *Allacma fusca*
3-3.5 mm long. Very common on bark, especially on that of beech trees, and also occurs on fungi.

PSOCOPTERA : Book Lice

27 Book Louse, *Psocus nebulosus*
6-7 mm long in the body. Very common, especially late in the summer, on the bark of various trees where it feeds on lower plants, including fungi. Both sexes are winged.

28 Book Louse, *Mesopsocus unipunctatis*
5 mm long. Lives in a similar way to the foregoing and is likewise very common. The female (28a) is wingless.

ORTHOPTERA : Cockroaches

29 Cockroach, *Ectobius lapponicus*
7-10 mm long. This cockroach occurs on plants and in withered leaves both in and out of woods. The female has only rudimentary wings whereas those of the male extend beyond the point of the abdomen.

30 Wood Cockroach, *E. lividus*
8-10 mm long. Both sexes have wings, but the males are very rare. Particularly plentiful on raspberry bushes.

DERMAPTERA : Earwigs

31 Wood Earwig, *Chelidurella acanthopygia*
The male may be up to 18 mm and the female (31a) up to 13 mm long.

Common in many European woods, living on trees and bushes. Both sexes are wingless.

ORTHOPTERA : Grasshoppers

32 Field Grasshopper, *Chorthippus* sp.

Male, 12-16 mm, female 15-22 mm long. Most grasshoppers are associated with open country but the genus shown here is also common in sunny open places in woods, including woodland paths. Field grasshoppers are distinguished from bush crickets by having shorter antennae, the females being without the sword-shaped ovipositor, the males stridulating by rubbing their hind legs against the covering wings, and the hearing organ being situated on the first segment of the abdomen.

ORTHOPTERA : Bush Crickets

33 Bush Cricket, *Meconema varium*

12-15 mm long. A European species which lives in the top of trees, especially oak, and is normally to be seen only after heavy storms or in the autumn. The male drums with his hind legs on the leaf upon which he is sitting. As is usual in the case of bush crickets, the hearing organ is situated in the forelegs immediately below the 'knee'.

34 Bush Cricket, *Pholidoptera griseoaptera*

Male 13 mm, female 18 mm long. Both have rudimentary wings. Common in July-September on bushes and low plants on the fringe of woods and in scrub. Often seen in gardens. Although the wings are short, the male is able to use them for stridulating.

35 Great Green Grasshopper, *Tettigonia viridissima*

Male 28-33 mm, female 32-35 mm long. Both sexes have full-grown wings. A carnivore, feeding on other insects. Mainly nocturnal: in the daytime, it generally remains motionless among the leaves and is then difficult to detect. The female is easier to see when she comes down to the ground to oviposit in the soil. The eggs over-winter and hatch in spring.

HEMIPTERA-HOMOPTERA: Plant Bugs

36 *Cixius nervosus*

6-8 mm long. Common on bushes and high herbs along the edges of woods and in clearings, especially in situations close to bogs. Like its relatives it sucks the juice of plants by inserting its proboscis into the tissues. A strong jumper.

37 *Centrotus cornutus*

7-8 mm long. Common on bushes and on low broad-leaved trees from May to July.

38 *Ledra aurita*

13-17 mm long. Widespread but not common. The adult animals are met with in July and August on hazel and alder. More sluggish than many related forms, and camouflaged by its resemblance to bark.

39 *Aphrophora salicis*

9-11 mm long. Very common from July to September on various trees where the nymphs, earlier in the summer, form 'cuckoo spit' (39a) consisting of a liquid secretion mixed with air bubbles. The nymph (39b) has the point of its abdomen, where its breathing hole is situated, in the surface of the froth.

40 *Philaenus spumarius*
5-6 mm long and very variable in colour and marking. Extremely common during the summer on grasses and other herbs both in woods and elsewhere. The nymph of the present species also produces cuckoo spit.

41 *Jassus lanio*
6-8 mm long. Common from July to September on oak and in oak scrub, especially on damp ground.

42 *Bythoscopus flavicollis*
5 mm long. Very variable in colour marking. Common on almost any species of broad-leaved tree throughout the summer.

HEMIPTERA-HOMOPTERA: Psyllids, Aphids and Coccids

43 **Jumping Plant Louse,** *Psylla alni*
3-4 mm long. Common on alder leaves. Psyllids are small jumping insects which, when adult, resemble cicadas and whose nymphs (43a) resemble aphids in that, like the latter, they suck the juices of plants. Such nymphs may be covered by a 'wool' (43b) consisting of fine threads of wax secreted by glands in the skin.

44 **Oakleaf Aphid,** *Phylloxera quercus*
About 1 mm long. Very common on oak leaves where its sucking action gives rise to small yellow spots (632). Underneath a spot of this kind the wingless female may often be found surrounded by eggs or newly hatched young. Winged aphids appear late in the summer.

45 **Aphid,** *Lachnus exsiccator*
4 mm long. Common on branches and young trunks of beech where its feeding activities produces cracks in the bark. A similar aphid, *L. roboris*, occurs on oak.

46 **Aphid,** *Adelges abietis*
1.6-2 mm long. This aphid lives partly on larch and partly on fir, where it causes 'pineapple galls' (577). The life history is complicated. The young of wingless females feeding on larch grow into winged females which fly to fir, where they lay their eggs at the bases of the needles. From these eggs come wingless males and females which suck the needles. These mate and give rise to young which remain to hibernate. In the following spring each of the young sucks upon its own leaf-base, and causes it to swell, the agglomerated swellings forming a gall resembling a diminutive pineapple. All these young aphids grow into females which lay eggs in the gall. From these eggs come insects which develop into winged females and which, in July and August, leave the gall. Some of them remain on the fir while the rest fly to a larch tree, where they lay eggs. The young from these eggs remain through the winter in cracks of the bark and next spring mature into wingless females, thus completing the life-cycle.

47 **Coccid,** *Cryptococcus fagi*
Barely 1 mm long. Small white woolly tufts consisting of very fine wax threads may often be seen on beech bark. Inside and underneath the 'wool' is the coccid (47a) a limbless lens-shaped animal, with a long proboscis which is inserted into the bark. All are females. Their eggs may be found late in the summer in the 'wool'. The larva is reddish in colour and has legs. It leaves the mother, inserts its proboscis into the plant tissues, and begins to form 'wool'.

48 Coccid, *Lepidosaphes ulmi*
The shield is 2-4 mm long and consists of two small patches at one end which are shed larval skins. The female coccid is partly covered by this shield and, in autumn, a hollow beneath the shield becomes filled with eggs which hatch the following spring. The small larvae have legs and disperse themselves over the trees, adhering to them and forming a new shield. They live upon various broad-leaved species, including fruit trees.

49 Coccid, *Chionaspis salicis*
The shield is 1.5 to 2.3 mm long on the female and about 1 mm long on the male. The freely moving larvae are carmine red. Common on willow and poplar and also found on ash, rowan, and alder.

**HEMIPTERA-HETEROPTERA:
Shield-bugs**

50 Green Shield-bug, *Palomena prasina*
12-15 mm long. Common from the middle of June in woods and gardens. Frequently found on the inflorescences of umbelliferous plants.

51 Sloebug, *Dolycoris baccarum*
10-12 mm long. Common throughout the summer on many different plants. Frequent on raspberries, to which it imparts an unpleasant taste from a secretion of its stink glands.

52 Birch Bug, *Elasmucha grisea*
7-9 mm long. Often seen in large numbers on birch. The female lays eggs on the birch leaf and at first guards them (52a), but takes no notice of the nymphs when these emerge.

53 *Troilus luridus*
11-13 mm long. Common in many woods where the grown bugs are to be found from the end of June onwards. The nymph (53a) differs greatly from the adult.

54 Common Shield-bug, *Pentatoma rufipes*
13-15 mm long. Commonly met with from mid-summer on various trees, especially on birch and alder.

55 *Picromerus bidens*
13-15 mm long. Easily recognised by the two spikes projecting from its thorax. Common in woods, where the fully developed bugs may be seen from late in the summer onward.

**HEMIPTERA-HETEROPTERA:
Capsid and Assassin Bugs**

56 Capsid Bug, *Stygnocoris rusticus*
3.5-4 mm long. Very common on withered hardwood leaves on the forest floor, especially in damp situations.

57 Capsid Bug, *Gastrodes abietum*
5-7 mm long. Often abundant in coniferous woods, especially underneath loose bark. Many over-winter in fallen cones.

58 Capsid Bug, *Aradus depressus*
5-6 mm long. Very common beneath loose bark, or in fungi growing on trees.

59 Capsid Bug, *Phytocoris populi*
6-7 mm long. Common on tree trunks in broad-leaved forests. Is believed to feed upon various small animals, such as aphids and insect larvae.

60 Capsid Bug, *Phylus coryli*
About 5 mm long. Common in deciduous woods where it occurs almost exclusively on hazel.

61 Capsid Bug, *Lygus pratensis*

4.5-7.5 mm long. Very variable in size and colour. A particularly common bug, to be found almost everywhere, in woods and open country alike.

62 Assassin Bug, *Empicoris vagabundus*

6-7 mm long. Easily recognised by the thin thread-like legs and antennae. Very common on shaded trunks and bushes. Walks slowly and awkwardly but flies well. A predator upon aphids and other small insects.

NEUROPTERA : Lacewings, Snake-Flies and Scorpion Flies

63 Ant-lion, *Myrmeleon formicarius*

Wing span about 35 mm. Somewhat resembles other lacewings, but the flight is awkward and occurs mainly at night. On the wing in June and July. The name 'ant lion' is derived from the larva (63a) which preys on ants caught in the funnel-shaped trap (63b) which the insect digs in the sand. The larva stays at the bottom of the trap with only its long, toothed mandibles projecting. If an ant comes to the edge of the trap the larva causes it to fall in by suddenly throwing sand at it with its head. The Ant-lion is unknown in Britain and is not general over much of the Continent, but occurs mainly in plantations on sandy ground.

64 Dead-leaf Lacewing, *Drepanopteryx phalaenoides*

Wing span about 12-15 mm. On the Continent is generally found in deciduous forests from April to October, but is frequently overlooked as it lives with its wings folded together with a remarkable likeness to a dead leaf. The larva (64a) feeds on aphids.

65 Common Goldeneye Lacewing, *Chrysopa vulgaris*

Wing span 12-14 mm. Very general from May to October both in broadleaved and coniferous woods. Often over-winters in houses. Each egg projects from the tip of a long stalk adhering to a plant (65a). The larva (65b) is about 7 mm long and feeds on the body-juices of aphids, which it sucks through hollow mandibles.

66 Hop Lacewing, *Hemerobius humuli*

Wing span 6-8.5 mm. Very general throughout the winter in deciduous woods and thickets. The eggs are laid singly on leaves. The larva (66a) reaches a length of about 7 mm and destroys aphids.

67 Snake-fly, *Rhaphidia xanthostigma*

Wing span 7-10 mm. All snake-flies are essentially woodland insects. There are four British species. The one shown in the figure is common from May to July in many continental pine woods on sandy soil. The name is derived from the extended first thoracic segment which resembles a neck. The female inserts her eggs in cracks in bark through a long oviduct. The larva (67a) lives underneath loose bark and preys upon various soft-skinned insects.

68 Common Scorpion Fly, *Panorpa communis*

Wing span 12-15 mm. Very common in June-August on bushes and herbs in woods of many kinds. So named from the development in the adult male (68a) of a terminal segment swollen and bent upwards like the sting of a scorpion. The larva (68b) has a specialised appendage at the apex of the abdomen whereby it secures itself to a support. It lives in the soil and is apparently omnivorous.

69 Snow Flea, *Boreus hyemalis*
3-5 mm long. The female (69a) is the longer as it has an ovipositor at the tip of the abdomen. Widespread in many European woods but often overlooked, especially as it generally leads a cryptozoic existence in winter underneath moss on stumps and trunks. Can sometimes be seen on sunlit snow.

MICROLEPIDOPTERA: 'Small' Moths

70 Small Ermine Moth, *Hyponomeuta euonymella*
Wing span 10-11 mm. *Hyponomeuta* includes a number of large moths whose larvae (181) live in a large communal web on various trees and bushes. The adults appear late in the summer and lay their eggs together on a branch. Here the eggs survive the winter, and hatch in the following spring.

71 Long-horned Moth, *Adela degeeriella*
Wing span 6-8 mm. Common in summer on raspberries, stinging nettles, etc., in woods.

72 Ash Moth, *Praus curticellus*
Wing span 7-8 mm. There are two generations, the first in June and the second late in August. The larva (684) sometimes causes considerable damage to young ash trees.

73 *Chimabache fagella*
Wings 10-11 mm. A characteristic defoliator of beech.

74 *Nepticula basalella*
Wings 3 mm. A common moth on

beech. The larva causes mines in the leaves (621).

75 *Tischeria complanella*
Wing span 3.5-4 mm. Common in oak woods where the caterpillars eat into the mesophyll of the oak leaves (631).

76 *Gracilaria syringella*
Wing span 5-6 mm. Associated both with ash and lilac. The feeding activities of the larvae produce discolouration and withering of the leaves (683).

77 *Lithocolletis faginella*
Wing span 4.5-5 mm. Often abundant in beech woods. The larvae make characteristic mines between the lateral veins of beech leaves (620).

78 *Acrobasis consociella*
Wing span 9-10 mm. Common in July in oak woods. The caterpillars spin the newly expanded oak leaves together before devouring them (186).

79 *Dioryctria abietella*
Wing span 12-15 mm. Common in coniferous woods in June-August, where the larvae frequently cause considerable damage (580).

80 *Crambus pratellus*
Wing span 9-10.5 mm. Very common in damp hardwood forests in June-July. The larvae live among the roots of water-logged vegetation.

81 *Evertia buoliana*
Wing 8.5-10.5 mm. Very common in fir plantations, where it can become a noxious pest (585).

82 Green Tortrix, *Tortrix viridana*
Wing span 8.5-11 mm. Very common on oak in woods during June-July

sometimes occurring in such numbers that the larva becomes a major defoliator (185).

83 Many-plume Moth, *Orneodes hexadactyla*

Wing span 7.5-8.5 mm. Common in woods where *Caprifolium* occurs in the shrub layer. Belongs with the next species to the feather-moths whose members have subdivided wings.

84 White-feather Moth, *Alucita pentadactyla*

Wing span 14-15 mm. The wings are divided on each side into five 'feathers'. When at rest these are held together as seen in the illustration. Very common both within and outside woods in June and July. The larva feeds on bindweed.

85 Bagworm, *Fumea casta*

Wing span 5-7 mm. Common in June-July. Only the female has wings. The larva (183) constructs a cocoon of grass, straw or similar material within which it pupates.

LEPIDOPTERA : Cossids, Clearwings, Swifts and Limacodids

86 Leopard Moth, *Zeuzera pyrina*

Wing span 20-35 mm. The female (86a) is much larger than the male. Widespread but not common. Flies in July-August. Persists for two years as a larva, which tunnels into various hardwood trees such as apple and ash.

87 Goat Moth, *Cossus cossus*

Wing span 30-40 mm. Flies in June-July and is very common in many parts of the country where it seems to be associated particularly with roadside poplars. Larva, 187.

88 Gold Swift, *Hepialus hecta*

Wing span 13-15 mm. The male is more strongly marked than the female (88a). Flies in June-July, when the males swarm in the evening. The larvae live mainly in the roots of ferns.

89 Festoon, *Apoda avellana*

Wing span 9-13 mm. On the wing in June-July. The female flies at night, the male also during the daytime. Uncommon. Larva, 188.

90 Raspberry Clearwing, *Bembecia hylaeiformis*

Wing span 10-13 mm. This, and the following species of clearwings, bear some resemblance to bees or wasps. Adults are about in July and August. Occur where raspberries grow, the larvae feeding in the roots and, before pupation, boring upwards into old stalks.

91 Large Red-belted Clearwing, *Aegeria culiciformis*

Wing span 9-12 mm. Flies in June to July. The larvae burrow into birch and alder immediately under the bark, often in fallen trees and stumps.

92 Hornet Moth, *Sesia apiformis*

Wing span 15-20 mm. The moth, which resembles a hornet, is seen from the end of June till late in July but flies little. Found in old poplars and, more rarely, in willow trees. The larva is found underneath the bark through which it bites a hole before pupating. The chrysalis has projections on its back and hind end whereby it pushes itself into this hole through which the adult moth escapes. The empty skin of the chrysalis (92a) remains in the hole after emergence.

LEPIDOPTERA:
Geometrid Moths

93 Chimney-Sweeper, *Odezia atrata*

Wing span 12-14 mm. Generally common around ditches and paths through woods. On the wing in June and July. The larva lives on wild chervil and other umbelliferous plants.

94 Large Emerald, *Hipparchus papilionaria*

Wing span 20-28 mm, being one of the larger woodland moths. Flies in July-August and is common in birth and alder woods. Larva, 189.

95 Clouded Magpie, *Abraxas sylvata*

Wing span 20-22 mm. Widespread and locally numerous. The adults are about in June and July and often spend the daylight hours at rest on a leaf, where they may resemble patches of bird excreta. Larva, 192.

96 Clouded Border, *Lomaspilis marginata*

Wing span 11-15 mm. Widespread throughout the country, flying most of the summer. The larvae occur mainly on aspen and willow and sometimes on hazel.

97 November Moth, *Oporinia dilutata*

Wing span 15-20 mm. On the wing between September and November. The eggs are laid in winter. The larvae may be found on several different trees, particularly oak, elm and apple.

98 Light Emerald, *Campaea magaritata*

Wing span 20-25 mm. Common in broad-leaved woods. The larvae may be found on beech, oak, hawthorn and hazel and the adult insect flies in June-July.

99 Single Dotted Wave, *Sterrha dimidiata*

Wing span 9-11 mm. A widespread species, usually with two generations in the year. The first of these flies in June, the second in August. The larva lives on various fir fungi.

100 Scallop Shell, *Calocalpe undulata*

Wing spang 15-18 mm. Locally common. Flies in June-July. The larva is mainly a defoliator of willow and poplar, but may also be found on bilberry, bog whortleberry and cowberry.

101 Winter Moth (Evesham Moth) *Operophtera brumata*

Wing span in the male 14-16 mm. The female (101a) has only vestigial wings. A common, generally distributed species in woods, timbered gardens, orchards, etc. The adult insects are about from October to December. Larva, 190.

102 August Thorn, *Ennomos quercinaria*

Wing span 19-22 mm. Colour is very variable, ranging from yellow ochre to brownish purple. Flies in August to September. Common in deciduous woods, the larva feeding on beech, oak, birch and hawthorn.

103 Early Thorn, *Selenia bilunaria*

Wing span 18-25 mm. Adults of the first generation fly in May, being followed in August by those of the second brood which are lighter in colour and smaller, with a wing span of 15-18 mm. Not rare in deciduous woods. Larva, 193.

104 Brimstone, *Opisthograptis luteolata*

Wing span 16-20 mm. General in luxuriant hardwood forest. Flies in

May-July, sometimes with a second generation in August-September. The larvae devour the foliage of many broad-leaved species.

105 Common White Wave, *Cabera pusaria*
Wing span 14-17 mm. Common in birch and alder woods. Flies in May-June and a new generation in July-August. The larva lives on birch and alder.

106 Mottled Beauty, *Boarmia repandata*
Wing span 20-24 mm. Common in many districts. Flies in June-July. The larva feeds on many different bushes and trees, ranging from bilberry and heather to willow and larch.

107 Peppered Moth, *Biston betularia*
Wing span in the male insect 20-23 mm, the female 25-30 mm. 107 shows the usual colour, 107a the melanistic form *carbonaria* which has become widespread in recent years. Common in many woods and timbered gardens. On the wing in May-June but the female flies little. Larva, 194.

108 Mottled Umber, *Erannis defoliaria*
Wing span in the male 20-24 mm. The female (108a) is wingless. Common. Often harmful, as the larvae (191) sometimes defoliate trees more or less completely. The adults are about from October to November.

109 Bordered White, *Bupalus piniaria*
Wing span 17-21 mm. The male and female (109a) differ in colour and are about between June and August. Common in pine woods throughout Britain, where the larvae (195) sometimes become pests.

LEPIDOPTERA : Noctuid and Lymantrid Moths

110 Grey Dagger, *Apatele psi*
Wing span 16-21 mm. Takes its name from the resemblance of the black markings on its back to the Greek letter 'psi'. Flies between late May and July. Often seen during the daytime on tree stumps and on poles. Common in hardwood forests and orchards. Larva, 197.

111 Nut-tree Tussock, *Colocasia coryli*
Wing span 13-17 mm. Common in hardwood forests where it is to be seen from May to June, followed, in many localities by a new generation in September. Larva, 196.

112 Herald, *Scoliopteryx libatrix*
Wing span 19-23 mm. Often hibernates in sheds and occupied dwellings. The larva is green with a yellow side-stripe and lives on willow and poplar. The adult may be encountered on willow catkins in early spring and is often still about in the late autumn.

113 Sprawler, *Brachionycha sphinx*
Wing span 17-20 mm. Flies later than most of the other noctuids, namely in October-November, and is characteristic of woodland in autumn. The larva, which is yellowish green with white longitudinal stripes, may be found on many different deciduous species.

114 Hebrew Character, *Orthosia gothica*
Wing span 15-18 mm. Common both in hardwood forests and open meadows from spring till early summer. The larva, which is green with three yellow stripes on its back and a wider white stripe on each side, feeds on different broad-leaved trees.

115 Merveille du Jour, *Griposia aprilina*

Wing span 19-22 mm. Although the name *aprilina* is derived from the fresh spring-green colour of the moth, the insect flies in the autumn, after the middle of September. Common in oak woods throughout the country. The larva is grey or brown with a black line along the side and is found mostly on oak where it lies concealed during the day in cracks of the bark.

116 Figure of Eight, *Episema caeruleocephala*

Wing span 16-20 mm. Very common in the autumn in hedgerows, scrub and gardens. The name *caeruleocephala* (blue head) refers to the larva (201)

117 Copper Underwing, *Amphipyra pyramidea*

Wing span 20-25 mm. Easily recognised by its size and its red hind wings. Locally common in woods and parks south of Oxfordshire. Flies in August. Larva, 199.

118 Large Yellow Underwing, *Tryphaena pronuba*

Wing span 23-27 mm. Very common throughout the summer. A fast-moving, diurnal moth, which is conspicuous in flight because of its brightly-coloured hind-wings, and then becomes hard to find when it drops suddenly to the ground and conceals them under its sombre forewings. Larva, 198. A closely related species *T. fimbria* has a much wider black band on the rear pair wings and is less active.

119 Red Underwing, *Catocala nupta*

Wing span 35-38 mm. Often common in south and east England where willows and poplars provide food for the larva. Flies in August-September and may sometimes be seen sitting upon tree trunks and poles in the daytime. Larva, 202.

LEPIDOPTERA : Noctuid, Thyatrid and Arctiid Moths

120 Clifden Nonpareil, *C. fraxini*

Wing span 40-48 mm and one of the larger noctuids. Formerly rare in parts of Europe, but during the last 30 years it has considerably extended its range and is now very common in many localities. Flies in August-September, mainly at night. The larva occurs mainly on the upper foliage of poplar and aspen.

121 Barred Sallow, *Tiliacea aurago*

Wing span 14-16 mm. Common in beech woods where it flies in September. The larva feeds upon beech; in its early stages living between leaves spun together, and when full-grown living free.

122 Snout, *Hypena proboscidalis*

Wing span 16-20 mm. The popular name 'Snout' is derived from the long feelers. Very common in woods where stinging-nettle grows, on which the larva (203) lives. There are usually two generations each year. The first of these flies from the middle of June until well into July, second in late August and September. More active during the day than most of the noctuids.

123 Orange Underwing, *Brephos parthenias*

Wing span 17-19 mm. Flies in the daytime in the early spring often high up in the tree tops. The larva feeds on birch.

124 Peach-blossom, *Thyatira batis*
Wing span 15-18 mm. A widespread
species. Flies from late in May to the
beginning of August in woods where
raspberries and blackberries grow,
which are the food-plants of the larva
(204).

125 Buff Arches, *Habrosyne derasa*
Wing span 16-19 mm. Locally common. Flies in June and July. The
reddish brown larva with round specks
on the side feeds on raspberries and
blackberries.

126 Pine Beauty, *Panolis flammea*
Wing span 14-17 mm. Widespread in
pine woods but relatively uncommon.
Flies in April and May. The larva
(200) resembles the needles of the
Scots pine (*Pinus sylvestris*).

127 Red-necked Footman, *Atolmis*
rubricollis
Wing span 14-16 mm. The name comes
from the red 'collar' which makes this
moth very easy to recognise. Widespread and sometimes common. Flies
in June and early July. The larva
devours lichens growing on fir, oak,
beech and sometimes on fences.

128 Common Footman, *Eilema*
lurideola
Wing span 15-17 mm. Common in
broadwood and pinewood forests, also
occasionally on heaths. Flies in July-
August. The larva, which is black with
a red-yellow stripe on its side, is
supposed usually to live on lichens
growing over branches and stones; but
it has been reared in captivity on the
leaves of such plants as sallow, apple
and oak.

LEPIDOPTERA : Tiger Moths

129 Scarlet Tiger, *Panaxia dominula*
Wing span 20-27 mm. Locally common in damp places in Britain
apparently only in parts of southern
England. Flies during July, often in
the daytime. The larva is black with
yellow spots and lives on many different
plants, chiefly herbaceous.

130 Wood Tiger, *Parasemia*
plantaginis
Wing span 16-21 mm. The colour
markings are very variable. Occurs
on moors and chalky hill-slopes and in
woods which are not thickly timbered.
Often about during the daytime in
June and early July. Larva, 206.

131 Ruby Tiger, *Phragmatobia*
fuliginosa
Wing span 13-18 mm. Very common
in many dry places, including woodland clearings, rough hillsides and
meadows. Usually there are two
generations in a year – the first of these
flies in May, the second in July and at
the beginning of August – but three
have been known. Caterpillars are
grey-brown in colour or nearly black,
without spots and having hair of the
same colour, and are to be found on
such herbs as dock, dandelion, golden-
rod (*Solidago*) and plantain.

132 Muslin Moth, *Diaphora mendica*
Wing span: male 11-13 mm, female
(132a) 14-18 mm. The two sexes are
also differently coloured. Widespread
and often common, and most likely to
be found in woods on poor soil. A
largely nocturnal species which flies in
May and June. The larvae are found
on herbs, such as plantain, dandelion,
dock and chickweed, and also woody
plants like birch and rose.

133 Garden Tiger, *Arctia caja*
Wing span 22-37 mm. Very variable in marking. Common in woods and gardens and also found in open situations. Flies in July and August only at night. The caterpillar (205) is popularly known as the 'woolly bear'.

LEPIDOPTERA : Prominents

134 Lobster Moth, *Stauropus fagi*
Wing span 22-27 mm. Widely distributed in beech woods but not often common. On the wing from May to July, at night. During the daytime it often rests low down on beech trunks and it then closely resembles a knot in the wood. The peculiar larva (207) has given rise to the popular name.

135 Coxcomb Prominent,
    Lophopteryx capucina
Wing span 16-20 mm. The commonest of the true Prominents. Plentiful in many beech woods and rather less so in other deciduous forests. In southern districts there are probably two generations in the year, one in May to June and the other in July to August. In the daytime the moth may be seen at rest near the bases of tree stumps. The larva (209) feeds chiefly on birch, beech and oak.

136 Swallow Prominent, *Pheosia tremula*
Wing span 20-25 mm. Widespread wherever there is a good growth of poplar and aspen. Occurs in two generations, the first at the end of May and beginning of June, the second the end of July and in August. Flies at night. Larva, 208.

LEPIDOPTERA : Miscellaneous Moths

137 Buff-tip, *Phalera bucephala*
Wing span 20-27 mm. The name is derived from the yellow spots on the fore wings. Common in hardwood forests, the caterpillar eating a wide range of tree foliage, particularly that of elm, lime and hazel. The moth flies at night in July and August, and during the day sits motionless with the wings rolled around its body, when it closely resembles a broken birch twig. Larva, 221.

138 Green Silver Lines, *Bena prasinana*
Wing span 15-18 mm. Easily recognised by its green wings and reddish antennae and legs. Common in many kinds of wood. The moth is on the wing between May and August, and is often to be seen in the daytime sitting on plants along woodland margins. The larva (210) feeds on oak, birch, beech and hazel.

139 Pebble Hook-tip, *Drepana falcataria*
Wing span 15-19 mm. Common in woods in damp situations. There are two generations, the first of which is greyer in colour and appears in May and June; the second, as illustrated, is on the wing in July and August. The green and brown larva lives in a web on the underside of birch and alder leaves.

140 Pale Tussock Moth, *Dasychira pudibunda*
Wing span 16-23 mm. Common in birchwoods throughout the country. Flies in May to June. During the day often sits on tree stumps in a position of rest with the forelegs extended. The larva (212) is striking.

141 Common Vapourer Moth,
    Orgyia antiqua
Wing span of the male 12-16 mm. The

female (141a) has no wings and, after emergence, remains on the surface of the cocoon. Here the male discovers and fertilisers her and she generally deposits her eggs on the cocoon where they over-winter. The male often flies in the daytime. The moth is about from late August to October, and is common in thickets and gardens. Larva, 213.

142 Black Arches, *Lymantria monacha*

Wing span of the male 18-20 mm, of the female (142a) 22-35 mm. In Britain, although widespread, the species is probably commonest in the New Forest. The caterpillar (214) feeds on the foliage of a variety of woodland trees, both broad-leaved and coniferous, sometimes in such numbers as to be a pest. The moth is about from July to September.

LEPIDOPTERA : Moths with Spinning Larvae

143 Oak Eggar, *Lasiocampa quercus*

Wing span of the male 25-27 mm, that of the female (143a), which is lighter coloured than the male, 32-35 mm. Generally distributed and often common. The name oak eggar refers only to the colour of the moth and the caterpillar probably never feeds on oak, although the food-plants are varied, ranging from bramble and hawthorn to ivy. Flies in July and early August. Larva, 217 .

144 Yellow-tail Moth, *Euproctis similis*

Wing span about 14-20 mm. As a rule the male is larger than the female and is without the black spot on the fore-wing. Common over most of the country in mixed broad-leaved forests, hedgerows and gardens. Flies in July-August. Larva, 216.

145 Lackey Moth, *Malacosoma neustria*

Wing span, male 15-17 mm, female 18-19 mm. The eggs are laid in a band around a small twig (215a). Common in undergrowth and in gardens. Flies in July-August. Larva, 215.

146 December Moth, *Poecilocampa populi*

Wing span, male 12-16 mm, female 16-18 mm. Very common in woods, timbered gardens and hedgerows. Flies in the autumn until December. The larva is grey with brown spots, is flatter than most of the other moth larvae, and is difficult to see when resting on a tree trunk. It devours the foliage of many broad-leaved species.

147 Fox Moth, *Macrothylacia rubi*

Wing span, male 20-33 mm, female 27-30 mm. The female is greyer. May be found in nearly all types of country. From the middle of May to the end of June, the male flies about at random in the daytime to look for the female, which rests close to the ground by day and flies only after dark. Larva, 218.

148 Pine-tree Lappet, *Dendrolimus pini*

Wing span, male 27-30 mm, female 32-42 mm. Very variable in colour and markings. In parts of the Continent, widespread in coniferous plantations, but there are few reliable records of its occurrence in Britain. Since both the adults and larvae keep to the tops of trees, they are seldom noticed. On the wing in July and early August. Larva, 219.

149 Small Eggar, *Eriogaster lanestris*

Wing span, male 12-16 mm, female 16-18 mm. Probably flies in April, but

F

the moth is seldom observed in nature. On the other hand, the characteristic dark brown caterpillars, with two rows of reddish yellow spots on the back, are often to be seen in May-July. They live in a large communal web on hawthorn, blackthorn, lime and several other species of trees. Generally distributed over much of southern England.

150 Saturnid, *Aglia tau*

Wing span, male 29-31 mm, female 38-43 mm. Unknown in Britain, but common in many beech woods on the Continent, where the male can be seen in full flight during May-June between the trees searching for the female which rests at the base of a trunk. Larva, 220.

LEPIDOPTERA : Hawk Moths

151 Eyed Hawk, *Smerinthus ocellata*

Wing span 30-38 mm. This and the following species are hawk-moths and are distinguished by the powerful torpedo shaped body and the long, narrow fore wings. All are excellent fliers. The eyed hawk is common throughout the southern half of England and is about from the end of May till into July. In favourable conditions there are two broods in the year. Flies at night and may sometimes be seen sitting on tree trunks or similar places in daytime. If disturbed it often shows the large spots by a sudden movement. Larva, 222.

152 Poplar Hawk, *Laothoe populi*

Wing span 35-45 mm. Varies greatly in colour from light grey to brown and reddish-yellow. Like the aforementioned it is associated with willow and poplar and may be met with throughout most of the summer. In the daytime it rests low down on tree trunks, and flies at night. The larva is similar to that of the eyed hawk but of a more golden green colour.

153 Pine Hawk, *Hyloicus pinastri*

Wing span 35-40 mm. In Britain widespread in pine woods, but not common. Usually on the wing from May to September. In the daytime it rests on trunks and in the evening sips nectar from honeysuckle and other tubular flowers. Larva, 221.

154 Elephant Hawk, *Deilephila elpenor*

Wing span 27-29 mm. Common in partly cleared woodland colonized by the rosebay, *Chamaenerion angustifolium*. Flies in June-July and may be seen shortly after sunset on lilac, honeysuckle and several other flowers. Larva, 223.

155 Lime Hawk, *Mimas tiliae*

Wing span 27-35 mm. Very variable in colour, it may also be green, red brown or a mixture of the two. Widely distributed and often common in part of southern England, in woods and parks containing lime trees. The larva which somewhat resembles that of the eyed hawk, has red or yellow slanting stripes and lives in the tops of lime trees.

156 Humming-bird Hawk, *Macroglossum stellatarum*

Wing span 20-22 mm. A native of the Mediterranean region, but a powerful long-range migrant which, in some years, is not rare in Britain where it may turn up anywhere. When it appears it often excites attention since it flies in the daytime and hovers with buzzing wings like a humming bird in front of such flowers as jasmine, from which it sucks nectar.

157 Broad-bordered Bee-hawk, *Hemaris fuciformis*

Wing span 18-20 mm. Its transparent wings are reminiscent of those of a bumble bee. It flies in morning sunshine during May, and June, when it sips nectar from such flowers as rhododendron and bugle. The larva which is green with a yellow side stripe and a red-brown horn, feeds on honeysuckle. Widely distributed and locally common in southern England.

LEPIDOPTERA : White and Nymphalid Butterflies

158 Black Apollo, *Parnassius mnemosyne*

Wing span 28-33 mm. Not found in British woods. On the Continent it flies in woodland clearings from the end of May to the beginning of July. The black larva with yellow spots lives on larch.

159 Green-veined White, *Pieris napi*

Wing span 23-25 mm. May be distinguished from the large and small white butterflies by the dark ribs on the underside of the wings (159a). Very common and more frequent in woods, especially in damp situations, than the other whites. There are two generations, the first of which flies in May-June and the second in August-September. Larva, 224.

160 Brimstone, *Gonepteryx rhamni*

Wing span 27-30 mm. The wings of the female are light green. Generally widespread, occurring especially at the edges of woods, where it flies in August-September and, after hibernating, again in early spring. Larva, 225.

161 White Admiral, *Limenitis sibylla*

Wing span 25-30 mm. Lives in damp woods but is not common, and seems to be confined in Britain to the south and east. A strong and graceful flier, on the wing in July. The larva is greenish yellow with red prickles and feeds on honeysuckle. It over-winters in a hibernaculum formed by drawing together the edges of a growing leaf and securing them with silk.

162 *Limenitis populi*

Wing span 35-45 mm. Not found in Britain. On the Continent, flies in July. The larva lives on aspen.

163 Purple Emperor, *Apatura iris*

Wing span 31-38 mm. An inhabitant of the larger oak woods in the southern and eastern counties of Britain. The male is frequently attracted to the juices of decaying animal carcases and excrement. The larva feeds on sallow and poplar.

164 Small Tortoiseshell, *Aglais urticae*

Wing span 25-27 mm. Abundant everywhere and specimens may be seen in practically every month of the year. There are two broods, one in June and the second in August-September. Individuals of the last brood hibernate as adults and reappear in early spring. The larva (226) feeds on stinging-nettle.

165 Peacock, *Vanessa io.*

Wing span 27-31 mm. Common in most regions. Two broods, the first in July, the second in early autumn. Adults of the second brood hibernate. As in the case of the small tortoiseshell, the larva (227) feeds on nettle, but caterpillars of the first brood generally appear later than the tortoiseshells.

166 Camberwell Beauty,
Euvanessa antiopa

Wing span 30-38 mm. Its occurrence in Britain is erratic, as it is essentially a migrant, probably from Northern Europe. Flies from the end of July into September. The caterpillar feeds mainly on sallow, willow and birch.

167 Glanville Fritillary, *Melitaea cinxia*

Wing span 18-22 mm. Only very local in Britain (chiefly in the Isle of Wight), but widespread on the Continent, where it is common along forest paths and the margins of woods in June. The larva, which is black with a reddish-brown head and limbs, is found on plantain, speedwell and hawkweed.

LEPIDOPTERA : Nymphalid, Satyrnid, Lycaenid and Pamphiline Butterflies

168 Silver-washed Fritillary, *Argynnis paphia*

Wing span 31-39 mm. Common in southern England and Wales where there are woods with violets in the herb-layer. Flies in July and a little into August. The larva bears imposing spines and generally feeds on dog violet (*Viola canina*).

169 Pearl-bordered Fritillary, *A. euphrosyne*

Wing span 18-21 mm. Fairly common in various broad-leaved woodlands, where it tends to frequent clearings. It generally flies in May-June, and sometimes as early as April. The larva (228) eats the leaves of violets.

170 Ringlet, *Aphantopus hyperanthus*

Wing span 20-22 mm. Common along forest paths and in woodland clearings. Flies June-July. The larvae, which are greyish with a dark stripe on the back and three whitish side stripes, feed on various grasses, particularly *Poa annua* and *Dactylis glomerata*.

171 Speckled Wood, *Pararge egeria*

Wing span 18-23 mm. Commoner in the south and west of England than in the east, and commoner still in Ireland. A butterfly of woodland rides and margins. There may be two generations, the first flying in May and June, the second in August and September. The green larva with white stripes feeds on various grasses including couch (*Agropyron repens*).

172 Small Heath, *Coenonympha pamphilus*

Wing span 13-17 mm. Very common both in and out of woods. Appears in two generations the first flying in May and June, the second August and September. The larva is dark green with a white double stripe on its back and a yellow side stripe, and lives on various grasses.

173 Green Hairstreak, *Callophrys rubi*

Wing span 12-15 mm. Often overlooked as the underside of the green wing (173a) makes the butterfly difficult to see when it is sitting on a leaf. Generally distributed in woodland clearings and similar places during May. The larvae feed on many different shrubs and herbs, including the petals of gorse (*Ulex*), the leaves of rock-rose (*Helianthemum*) and the berries of buckthorn (*Rhamnus*).

174 White-letter Hairstreak, *Thecla w-album*

Wing span 14-17 mm. Takes its name from the two white lines on the underside (174a). Widespread but not very conspicuous. The adult is on the wing in July-August, and generally flies around elm-trees. The larvae may be found on oak, lime and other deciduous trees.

175 Purple Hairstreak, *Zephyrus quercus*

Wing span 14-17 mm. Widespread in oak woods, but is often overlooked since both the adult and larva are usually confined to the crowns of the trees. Flies in July-August.

176 Small Copper, *Heodes phloeas*

Wing span 12-16 mm. Widespread and common both in and out of woods. The adult is about from April to October and there are two generations in the year. The larva feeds on dock and sorrel.

177 Common Blue, *Polyommatus icarus*

Wing span 12-16 mm. Male and female (177a) are very different. The commonest of the 'blues' in Britain. May be found in clearings in woods, along woodland margins and paths, near hedgerows, etc. The first brood flies in May-June and the second in August-September. Larva, 229.

178 Essex Skipper, *Adopaea lineola*

Wing span 11-13 mm. May be recognised by its wide head. Well distributed in south-east England, where it occurs along the edges of woods and in rough ground. Flies in July-August. Larva lives on various coarse grasses.

179 Large Skipper, *Augiades sylvanus*

Wing span 12-15 mm. (The female does not have the black patch on the forewing.) Common both in and near woodland. Flies in June-July. The larva lives in the blade of such a grass as cocksfoot (*Dactylis glomerata*), the edges of which are fastened together with silk.

LEPIDOPTERA : Miscellaneous Moth Larvae

180 *Incurvaria koerneriella*

The larva bites a patch out of an oak or beech leaf and uses this as a transportable case in which it falls to the ground. Here it begins to devour dead leaves. As it grows, it bites pieces out of withered leaves and adds these to its case, in which the larva over-winters and pupates.

181 Small Ermine Moth, *Hyponomeuta euonymella*

Up to about 20 mm long. Lives in a large communal web on bird cherry. Similar species are found on hawthorn, blackthorn and apple. Parasite, 447.

182 Bagworm, *Talaeporia tubulosa*

The bag is about 15 mm long. The larva spins it on trunks, stakes and similar places in May and pupates. The adult appears in June-July.

183 Bagworm, *Fumea casta*

The bag, covered with small pieces of straw, is 8-12 mm long and is very common on tree trunks in the spring. Adult, 85.

184 Bagworm, *Coleophora laricinella*

The larva lives at first by eating into larch needles. Towards autumn it constructs a shelter from the tips of the needles and over-winters here. When the new needles expand in spring, it resumes its feeding and completes its metamorphosis.

185 Green Tortrix, *Tortrix viridana*
The larva is up to 16 mm long, and feeds on oak leaves, from which it may often be seen hanging on a thread. When fully grown the larva bends a leaf into a roll and secures it by silk. Pupation takes place in the roll. Adult, 82.

186 *Acrobasis consociella*
The larva is about 15 mm long, and is common in oakwoods in June. It spins oak leaves together. Adult, 78.

187 Goat Moth, *Cossus cossus*
Up to 100 mm long, this impressive-looking caterpillar has a sour, 'goaty' smell. Common in the trunks of several broad-leaved species, especially willows and poplars. In cases of severe infestation, the long tunnels which it eats into the wood may so weaken the tree as to cause its collapse in a storm. The larva, which takes several years to mature, pupates in the ground. Adult, 87.

188 Festoon, *Apoda avellana*
12-13 mm long, this flat, louse-like caterpillar feeds on oak and beech, and bears little resemblance to the larva of a moth. It may be found during August but is not common. Adult, 89.

LEPIDOPTERA : Geometrid Moth Larvae

189 Large Emerald, *Hipparchus papilionaria*
32 mm long. Hibernates when small. The grown larva is found in May and June, especially on birch but also on hazel and elm on the leaves of which it feeds. The larva moves with the 'looping' action typical of the group. Adult, 94.

190 Winter Moth (Evesham Moth), *Operophtera brumata*
20 mm long. The eggs laid in the autumn on branches of fruit trees, etc., hatch in May. At first the small larvae eat their way into the buds, later they spin the expanding leaves together and eat them. If the infestation is severe, the larvae may completely strip a tree by the middle of June. Pupation occurs in the soil and the pupa lies quiescent for about five months before the moth (101) emerges.

191 Mottled Umber, *Erannis defoliaria*
32 mm long. Life history similar to the foregoing. Adult, 108.

192 Clouded Magpie, *Abraxas sylvata.*
32 mm long. This easily recognisable larva feeds mainly on wych elm. Adult, 95.

193 Early Thorn, *Selenia bilunaria*
30-34 mm long. Recognisable by the double pointed hump on the seventh and eighth segments. May be seen in the summer on various trees and bushes, including lime, willow, elm and rose, and, when resting, is very like a twig. Adult, 103.

194 Peppered Moth, *Biston betularia*
55-58 mm. The larva may be grey brown or green. When disturbed it extends the body at an angle of 45° from its support while holding fast with the prolegs and, in this position, it resembles a small branch. It colonizes various hardwood trees, especially birch, but may also be seen on raspberry. Adult, 107.

195 Bordered White, *Bupalus piniaria*
17-21 mm long. Feeds from August to October on the needles of Scots pine and other firs. It is sometimes a serious pest. Adult, 109.

LEPIDOPTERA : Noctuid Moth Larvae

196 Nut-tree Tussock, *Colocasia coryli*
30-40 mm long. May be found in the autumn on birch, hazel and hornbeam and especially on beech, where it may be very numerous. It tends to colonise trees in somewhat open situations. Adult, 111.

197 Grey Dagger, *Apatele psi*
30-40 mm long. A diagnostic feature is the prominent black hair-tuft on segment four. Common in the autumn on many deciduous trees including fruit trees in gardens. Adult, 110.

198 Large Yellow Underwing, *Tryphaena pronuba*
Up to 50 mm long. May also be yellowish or greenish. Common from August on many different herbaceous plants and sometimes a pest in vegetable gardens. Adult, 118.

199 Copper Underwing, *Amphipyra pyramidea*
40-46 mm long. The specific name *pyramidea* is derived from the pointed protuberance on the eleventh segment. Locally common from April to June on such broad-leaved species as birch, sallow, oak, plum and rose. Adult, 117.

200 Pine Beauty, *Panolis flammea*
33 mm long. Resembles the pine-needles on which it feeds, and sometimes occurs in such numbers as to be a pest. Found in the summer on fir trees. Adult, 126.

201 Figure of Eight, *Episema caeruleocephala*
40 mm long. Very common in early summer on blackthorn, hawthorn and hazel, in open woods, hedgerows, etc., and often on fruit trees in gardens, where it may be harmful. Note the bluish head (=*caeruleocephala*). Adult, 116.

202 Red Underwing, *Catocala nupta*
About 60 mm long. The caterpillar feeds at night between April and July on willows and poplars, and may be found in bark crevices during the day-time. It is not easy to locate. Adult, 119.

LEPIDOPTERA : Spinning Larvae of Moths

203 Snout, *Hypena proboscidalis*
25 mm long. Common on stinging nettles and sometimes on hops. Adult, 122.

204 Peach-blossom, *Thyatira batis*
35-40 mm long. Easily recognised by its pink colour and the cowl-like hump on the second segment. It feeds in autumn on raspberry and blackberry leaves. Adult, 124.

205 Garden Tiger, *Arctia caja*
About 60 mm long. The 'woolly bear' caterpillar. It over-winters while still small and is met with in the spring and early summer on the many different herbaceous species which serve as food-plants. Before pupating it spins a large cocoon into which its long hairs are incorporated. Adult, 133.

206 Wood Tiger, *Parasemia plantaginis*
About 30 mm long. Over-winters when small and may be found in the spring on various plants, especially

forget-me-not (*Myosotis*) and plantains. Cocoon-spinning follows the general lines of the previous species. Adult, 130.

207 Lobster Moth, *Stauropus fagi*
About 60 mm long. This long-legged caterpillar, the colour of a boiled lobster, with its angular segments and which takes up such characteristic postures, cannot be confused with any other British species. It is to be seen in the late summer especially on beech, but sometimes also on oak, hazel, birch and other hardwood trees. Adult, 134.

208 Swallow Prominent, *Pheosia tremula*
50-60 mm. The larva had red prolegs and is generally light green, although it may also be brown. Seen mostly in the late summer on aspen or other poplars, and more rarely on willow. Adult, 136.

209 Coxcomb Prominent,
Lophopteryx capucina
About 50 mm long. Usually light green, but a reddish phase also occurs. The two scarlet 'warts' on segment eleven are diagnostic. When disturbed bends the fore part of its body backwards. Common between July and October on beech, oak, birch, etc. Adult, 135.

210 Green Silver Lines, *Bena prasinana*
24-28 mm long. Predominantly green and yellow. The prolegs tend to be splayed laterally and the anal claspers are tinged red. Common in the autumn on beech and several other hardwood trees. Adult, 138.

211 Buff-tip, *Phalera bucephala*
About 60 mm long. This very characteristic stoutly-built larva is found in the summer on various broad-leaved trees, often in communities. It pupates in the ground. Adult, 137.

212 Pale Tussock, *Dasychira pudibunda*
40-45 mm long and one of the most beautiful of our moth larvae. Pale green, bearing golden hair-tufts like diminutive shaving brushes, and an anal tuft of reddish hairs coming to a point. Found in the late summer and autumn on various hardwood trees especially beech. When present in large numbers, the larvae may strip all the leaves from a single tree. Adult, 140.

213 Common Vapourer, *Orgyia antiqua*
12-16 mm long. Common in the summer on many species of bushes and trees in woods and gardens. Although of a more smoky grey colour than the previous species, and smaller, its hairy tufts are equally remarkable. Adult, 141.

214 Black Arches, *Lymantria monacha*
35-40 mm long. The caterpillar is fully grown in July and occurs both in broad-leaves and coniferous woods where it sometimes causes considerable damage. The brownish pupa, which has a metallic lustre, is enclosed in a transparent cocoon fastened to bark. Adult, 142.

215 Lackey Moth, *Malacosoma neustria*
40-50 mm long. The blue and red larvae are found in early summer within a web on various broad-leaved trees and bushes, where they sometimes cause great damage. 215a shows the characteristic pattern in which the eggs are deposited. Adult, 145.

216 Yellowtail Moth, *Euproctis similis*

About 35 mm long. Found during the early summer in woods, hedgerows and gardens on a wide range of trees and shrubs. The conspicuous markings – vermillion, black and white – are apparently a form of warning colouration: the caterpillar is hairy and can induce an unpleasant skin-rash if handled. Adult, 144.

217 Oak Eggar, *Lasiocampa quercus*

70-80 mm long. Over-winters as a small larva, becoming full grown in May-June. Feeds on many different plants, both herbaceous and woody, but apparently never on oak. Adult, 143.

218 Fox Moth, *Macrothylacia rubi*

60-80 mm long. When young it is black with a yellow band. The caterpillar is active from July until well into the autumn, hibernating when nearly full grown, and pupating in the spring. Food-plants include blackberry and raspberry. Adult, 147.

219 Pine-tree Lappet, *Dendrolimus pini*

Up to 100 mm long. Occurs in June-July in coniferous plantations, usually near the tops of the trees. Apparently very rare in Britain: in Northern Germany and in Norway it has sometimes become a pest. Adult, 148.

220 Saturnid, *Aglia tau*

50-65 mm long. Unknown in Britain. In continental woods, the grown larva is seen in July-August on beech or more rarely, on birch and oak. Adult, 150.

LEPIDOPTERA : Larvae of Hawk Moths

221 Pine Hawk, *Hyloicus pinastri*

Up to 90 mm long. The terminal horn is characteristic of the hawk-moth larvae. Widespread, but not common, from July to September on Scots pine and other firs. Adult, 153.

222 Eyed Hawk, *Smerinthus ocellata*

Up to 90 mm long. Found from July until the autumn on willow, poplar, blackthorn and fruit trees. Adult, 151.

223 Elephant Hawk, *Deilephila elpenor*

70-80 mm long. Easily recognised by the swollen anterior end bearing white spots, giving the impression of a face with staring eyes. Met with in August-September on willow herb and bedstraw.

LEPIDOPTERA : Larvae of Butterflies

224 Green-veined White, *Pieris napi*

30 mm long. Lives on various crucifers and may occur during most of the summer. The pupa (224a) is attached to its support by the point on the abdomen, and by a medium girdle spun around the body. Adult, 159.

225 Brimstone, *Gonepteryx rhamni*

About 50 mm long. Found from May to June on common buckthorn. The chrysalis (225a) is attached in the same way as the foregoing. Adult, 160.

226 Small Tortoiseshell, *Aglais urticae*

About 40 mm long. Gregarious on stinging nettle in July-September. The pupa (226a) hangs loosely downwards attached only by its hind end, which often is surrounded by the remains of the last larval skin. Adult, 164.

227 Peacock, *Vanessa io*
40-45 mm long. Gregarious on stinging nettle, less commonly on hop, in July-August. The chrysalis (227a) hangs loosely down. Adult, 165.

228 Pearl-bordered Fritillary, *Argynnis euphrosyne*
About 30 mm long. Found in June-July on violet, and sometimes on whortleberry. Adult, 169.

229 Common Blue, *Polyommatus icarus*
18-25 mm long. Found in summer on various papilionaceous flowers, such as rest harrow and clover. It over-winters in the pupal state. Adult, 177.

COLEOPTERA : Ground Beetles

230 *Calosoma inquisitor*
16-21 mm long. Widespread, but not common. Noticed particularly in oak woods in June, where it may be found on trees and bushes searching for the larvae and pupae of other insects. Like other ground beetles, the species is largely predatory.

231 *Carabus coriaceus*
34-40 mm long and the largest of the northern European ground beetles. Common, especially in spring and autumn, in broad-leaved woods, where it hunts on the ground during the night. Like other ground beetles it has stink glands in its abdomen.

232 *C. hortensis*
23-28 mm long. Common in woods of several different types, especially in August and September, when it hunts at night for worms and larvae.

233 Violet Ground Beetle, *C. violaceus*
22-29 mm long. Common in woods, more especially those on relatively dry ground, and most frequently observed between June and August. Larva, 371.

234 *Cychrus caraboides* var. *rostratus*
16-21 mm long. May be found almost throughout the whole year underneath the leaves, moss and bark in broad-leaved woods. It lives almost exclusively on snails. Larva, 372.

235 *Feronia niger*
15-21 mm long. Very common from April to August on damp ground in more open-canopied woods. Also often encountered in gardens and cellars.

236 *Platynus assimilis*
10-12 mm long. Very common underneath moss and leaf litter in woodlands, especially during spring and autumn.

237 *Nebria brevicollis*
10-13 mm long. Common on rather damp ground in broad-leaved forests. May be found during the whole of the year but occurs most abundantly in July-August.

238 *Anax parallelopipedus*
18-22 mm long. Common, especially in spring, on damp ground in such broad-leaved forests as beech woods.

239 *Dromius quadrimaculatus*
5-6 mm long. Common throughout most of the year, occurring particularly under the bark on dead branches, both of broad-leaved and coniferous trees.

240 *Ophonus seladon*
8 mm long. Common in the summer on partly shaded ground both in woods themselves and beyond.

241   *Harpalus latus*
7-9 mm long. Common the whole of the year, but most plentiful in the summer on the woodland floor underneath moss, fallen branches, among litter, etc.

242   Tiger Beetle, *Cincindela campestris*
12-15 mm long. Common in May in open, sunlit situations, and may be found both in woodland clearings and beyond. Moves very fast and often flies. Larva, 373.

COLEOPTERA : Carrion Beetles

243   *Proteinus brachypterus*
1.6-2 mm long. Particularly common in swampy places on dead animals and the like, especially in the spring and autumn. This and the following nine predatory species may be recognised from their short elytra.

244   *Gyrophaena affinis*
1.5-2 mm long. Very common between July and October, on many different kinds of woodland fungi.

245   *Atheta fungi*
2.2-2.8 mm long. Very common all over the country in moss, wood of deciduous trees, fungi, ant hills, and many other places. The *Atheta* genus includes numerous other species.

246   *Anthophagus caraboides*
4.5-5 mm long. Found from June to September on broad-leaved trees or bushes in woods or scrubland.

247   *Bolitobius lunulatus*
5-6 mm long. Common in swampy places, especially during the autumn.

248   *Conosoma testaceum*
4-4.5 mm long. Common in hollow trees, in rotten wood and in leaf-litter throughout the year.

249   *Tachyporus obtusus*
3.5-4 mm long. Very common in spring and autumn, on the woodland floor, and beyond it, in clearings and meadows.

250   Devil's Coach-horse, *Ocypus olens*
24-30 mm long. The largest of our predatory beetles. Common during the summer in ground under hardwood and moss. In adopting a characteristic attitude of defence, the insect elevates the tip of its abdomen and directs it forward.

251   *Staphylinus brunnipes*
12-14 mm long. Common from February to September in similar places to the foregoing, but also frequent in gravel pits.

252   *Othius punctulatus*
11-14 mm long. Common on the ground in woods and on carrion, manure and fungi.

253   *Catops picipes*
5-6 mm long. Widespread, but seems very rare. May be found in the spring and autumn in hollow trees and especially in the nests and passages made by mice.

254   Sexton Beetle, *Necrophorus humator*
18-25 mm long. Common especially on the carcasses of small birds and mammals. They feed on these and also bury them, afterwards laying their eggs underground, near the buried meat. The adults are about between April and October. Larva, 374.

255 Sexton Beetle, *N. investigator*
18-22 mm long and the commonest burying beetle in woods. Habits are similar to the foregoing but the adults are abroad only from June to September.

256 Sexton Beetle, *Necrodes litoralis*
15-25 mm long. Widespread both in woods and beyond. Sometimes to be seen in numbers on large carcasses. To be found from May to August.

257 Red-breasted Carrion Beetle,
Oeceoptoma thoracica
12-16 mm long. Common in woods May to September where it obtains its food from carcasses, rotting fungi, etc.

258 *Xylodrepa quadripunctata*
12-15 mm long. Very common in woods especially in oak woods, where it may be found from May to June on trees, hunting for the larvae of Lepidoptera.

259 Black Carrion Beetle, *Silpha carinata*
11-12 mm long. A widespread species which is often to be found on fungi. Its diet includes snails and worms. Is active in May to August.

260 *Hister striola*
5-7 mm long. Common in summer underneath bark, on stumps, and also often in the sap exuded from beech and birch trees, but less often on actual carrion.

COLEOPTERA: 'Softwing' Beetles

261 Glow-worm, *Lampyris noctiluca*
The male is 11-12 mm long, the similarly coloured female (261a) 16-18 mm. Widespread over most of the country, but not common. May be found especially in rough ground and at the edges of woods in June and July. The male swarms on warm nights looking for the flightless females which sit in the grass and are strongly luminous. The male, larva (376) and egg are all luminous, but less so than the female. The food consists of snails.

262 Soldier Beetle, *Cantharis fusca*
11-15 mm long. Common over the whole country in June to July, both in and out of woods, where it may be encountered in the inflorescences of herbaceous plants growing along the sides of paths. Larva, 377. This and the following species are sometimes termed 'softwing' beetles, on account of the thin and flexible elytra.

263 Soldier Beetle, *C. livida*
10-13.5 mm long. Like the foregoing

264 *Rhagonycha fulva*
7-10 mm long. Very common on flowers in June and July.

265 *Opilo mollis*
9-13 mm long. Widespread but not particularly common. Found mostly on the dead trunks and branches of various broad-leaved trees which have been attacked by boring or bark beetles, the larvae of which provide it with food.

266 Ant Beetle, *Thanasimus formicarius*
7-10 mm long. Commonly seen in pine woods, especially during the summer where it is active in the daytime and hunts for bark beetles, upon which it feeds both as a larva (385) and when full grown. Its manner of moving is reminiscent of a wood ant.

267 *Dasytes caeruleus*
5-7 mm long. During May and June

frequents flowers in woods and scrub-land. The larva lives on rotten wood.

268   *Malachius bipustulatus*
5-7 mm long. Common, especially during the summer, on the inflorescences of grasses and other plants, the pollen of which it devours.

269   *Malthodes marginatus*
4-4.5 mm long. Occurs in June and July on many broad-leaved trees and bushes. Often about on flowers also.

COLEOPTERA : Skipjack Beetles

270   *Chrysobothris affinis*
12-15 mm long. Elytra dark brown, with coppery depressions which are situated on the three raised striae and interrupt them. It occurs on stumps and the larva lives in oak and pine. Active from June to August. A native of various European countries, but not of Britain. Larva, 378.

271   *Agrilus viridis*
6-9 mm long. Very common in June-July in woods where they may be seen swarming in the mid-day sun over willow and hazel, and more rarely over other trees and bushes.

272   *Denticollis linearis*
9-13 mm long. Common in scrub and woods in May-July. Belongs with the following seven species to the skipjacks, so named on account of the jumps they make from a prone position, when lying on their backs. The leap is effected with the aid of a specialised apparatus on the posterior edge of the first segment, which can be retracted into a cavity in the anterior edge of the second segment with such force that the beetle is flung into the air. Larva, 379.

273   *Lacon murinus*
10-20 mm long. Common both in and out of woods. Often flies, in sunshine, during May-July.

274   *Agriotes aterrimus*
12-15 mm long. Frequents woods and scrub in June.

275   *A. acuminatus*
6-8 mm long. Very common during June, especially in oak woods.

276   *Elater cinnabarinus*
12-13 mm long. Widespread but not common. The larvae live on rotten beech trees and, less frequently, on poplar.

277   *Melanotus rufipes*
13-19 mm long. Widespread in woods. Flies at night in June. The larvae live on rotten timber both of broad-leaved and coniferous trees.

278   *Corymbites sjaelandicus*
10-16 mm long. Frequent on the edges of woods towards midsummer. The larvae live in the soil.

279   *Athous haemorrhoidalis*
9-16 mm long. Very common in woods and scrub in May to July where it may be often seen on flowers. The larvae live in the ground and may sometimes cause great damage in nurseries.

COLEOPTERA: Miscellaneous Beetles

280   **Cardinal Beetle,** *Pyrochroa coccinea*
14-15 mm long. Very common in June on the inflorescences of *Umbelliferae*, hawthorn, etc. Larva, 381.

281 *Metoecus paradoxus*
8-12 mm long. Widely distributed, including Scotland and Ireland, but rare. Breeds in the nest of the Common Wasp (*Vespa vulgaris*). The adult beetles are about in August and September.

282 *Oedemera femorata*
8-10 mm long. Widespread in deciduous forests, where it may be seen in June and July on flowering umbelliferous plants, notably on *Aegopodium podagraria*.

283 Spanish Fly, *Lytta vesicatoria*
12-21 mm long. Widespread, but seldom numerous, occurring sporadically on ash; and more rarely on elder, honeysuckle and other bushes. Seen in June and July. Also known as the 'Plaster Beetle'. It contains a substance which causes blisters, so that it was formerly used as a counter-irritant in medical plasters.

284 *Rhinosimus planirostris*
3-3.5 mm long. Common in spring and autumn in broad-leaved woodlands, where it may be found on dead branches and under loosened bark.

285 *Tetratoma fungorum*
4-4.5 mm long. Widespread in woodlands, where it may be seen in spring and autumn on stumps and trunks, and especially on fungi growing on conifers.

286 *Tomoxia biguttata*
6.5-8 mm long. Widespread over most of the country, especially in the southern part. To be seen in June-July on stumps and on flowers. Quick-moving and shy. (The hind end of the body is drawn out into a point bearing a superficial resemblance to a 'sting'.) Larva, 380.

287 Raspberry Beetle, *Byturus urbanus*
3.8-4.3 mm long. Common on flowers and raspberries in June-July. Eggs are laid in developing flower-buds. The larva (382) is the 'worm' in raspberries

288 *Nitidula bipunctata*
3-5 mm long. Common throughout the country in April to September. Both the adult and the larva live on wind-dried carcases and bones.

289 *Glischrochilus* var. *quadripunctatu.*
3-6 mm long. Very common from March to November under the bark o coniferous trees, where it often tunnels through the looser material.

290 *Rhizophagus dispar*
3-4 mm long. Common between April and November under the bark both o broad-leaved and of coniferous trees Often found inside holes made by bark beetles, whose larvae it devours.

291 *Bitoma crenata*
2.6-3.5 mm long. Resembles the foregoing, but seems to prefer beech.

292 *Cerylon histeroides*
1.8-3.5 mm long. Very common throughout the country underneath bark both of broad-leaved and coniferous trees. Believed to live on the larvae of bark beetles.

293 Fungus Beetle,
Mycetophagus quadripustulatus
5-6 mm long. Common throughout the summer season on various tree fungi especially those on ash and oak.

COLEOPTERA: Ladybirds
294 Ten-spot Ladybird,
Coccinella decempunctata
3.5-5 mm long. Varies greatly as

regards the distribution of colour on the elytra (294a and b). Common from May to October on trees and bushes. Belongs like the following species to the ladybird group which, both as adults and larvae, generally feed on aphids.

295 Fourteen-spot Ladybird,
C. quatuordecimpunctata
3.5-4.5 mm long. Recognisable from the rectangular spots. Very common throughout the summer on a wide range of plants.

296 *C. quatuordecimguttata*
4.5-6 mm long. Common on trees and bushes from May to October.

297 Seven-spot Ladybird,
C. septempunctata
5.5-8 mm long. One of the commonest ladybirds, which may be met with almost throughout the year. In the summer it may be present in thousands both in the adult and larval stages, on plants attacked by aphids; and in winter the adults may be found hibernating in sheltered situations, including the insides of dwelling-houses. Larva, 383.

298 Eyed Ladybird, *Anatis ocellata*
8-9 mm long and the largest ladybird in Britain. Very common on conifers.

299 *Paramysia oblongoguttata*
6-8 mm long. Very common on coniferous trees both in early and late summer.

COLEOPTERA: Boring Beetles

300 Boletus Beetle, *Cis boleti*
2.8-3.5 mm long. The commonest member of its genus in Britain. Found throughout much of the year in various woodland fungi, notably *Boletus*.

301 *Ernobius mollis*
3.5-5 mm long. Widespread from June to August in coniferous forests. Breeds in dead branches. Larva, 384.

302 Furniture Beetle,
Anobium punctatum
2.5-4 mm long. Commonest in July-August. Breeds in dead, dry wood of both broad-leaved and coniferous trees; and frequently enters houses, where it becomes a serious nuisance in colonising furniture and structural timber.

303 Furniture Beetle,
Ptilinus pectinicornis
3-5 mm long. The antennae of the male are very conspicuous and flabellate; those of the female smaller and pectinate. Very common in dry wood of old oaks, willows, alders, etc., and often seen in July on beech trunks from which some of the bark is missing. It also comes indoors and attacks furniture, but usually only that made from beech or maple.

304 *Hylecoetus dermestoides*
6-18 mm long. The male (304a) is smaller and darker than the female. This peculiar elongated beetle is of local distribution. The adult is usually most active between May and June. The larva (386), which is distinctive, colonises dead or dying trunks, or old stumps of broad-leaved trees.

COLEOPTERA : Scarabaeid Beetles

305 *Serica brunnea*
8-10 mm long. Belongs, like the next ten species, to the *Scarabaeidae*, which may be known by their characteristic clubbed antennae and generally by having powerful legs adapted for

digging. The present species is widespread in woods, plantations, and gardens, particularly those established on sandy soil.

306 Garden Chafer,
Phyllopertha horticola

8-12 mm long. It often swarms in June and July on sunlit trees – particularly fruit trees – the leaves and flowers of which it devours. The larva feeds on roots in the soil and takes two years to develop.

307 *Aphodius fimetarius*

5.5-7 mm long. Probably the commonest of the 41 British species of *Aphodius*, both the adults and the larvae of which feed on manure. The present species may be found throughout most of the year on manure of almost any kind.

308 *Sinodendron cylindricum*

12-16 mm long. Widely distributed and locally common, both as adults and larvae, in rotten beech trunks and stumps, but rare in other broad-leaved trees. The beetle swarms in the evening towards midsummer. Development takes two years.

309 *Systenocerus caraboides*

10-14 mm long. The male is narrower and bluer than the female. The larva occurs in stumps of beech and oak; and after emerging from the pupa in autumn, the adult beetle remains in the stump until the following spring, when it may be seen eating the buds and leaves of oak. It swarms in sunshine.

310 Greater Stag Beetle,
Lucanus cervus

Male up to 75 mm long, and female (310a) up to 50 mm long. The largest European beetle. In Britain, it seems to be mainly an insect of London and the Home Counties. Lives as a larva in old rotten oak where development takes five years. The adult beetles emerge late in June, and the males use their prodigious mandibles for fighting before mating with the females.

311 Lesser Stag Beetle,
Dorcus parallelopipedus

19-32 mm long. Widespread in rotten stumps and trunks of beech; rare in those of other broad-wood trees. Most frequently seen in late June. Larva, 387.

312 Bumble-dor,
Geotrupes stercorarius

14-20 mm long. Very common in open woods where it may often be seen on paths looking for manure which it buries underground and in which it lays its eggs. Often seen flying on warm summer evenings. One of its English names, 'Lousy Watchman', refers to the fact that it is commonly infested with mites (535).

313 Rose Chafer, *Cetonia aurata*

14-22 mm long. Common in sunlit parts of woods and their margins, and often seen in the flowers of elder, meadowsweet, rose, etc. Most frequent in July. The elytra are slightly curved towards the anterior end. The wings fit into the hollow, and there the beetle is able to fly with the elytra closed, by contrast with most other beetles whose elytra are raised during flight. Larva, 390.

314 Rhinoceros Beetle,
Oryctes nasicornis

The female does not have the horn on the head. Found in parts of Europe, but not in Britain, in sawmills and similar places where there are fermenting sawdust and wood shavings in which the larva (388) lives. The beetles swarm in summer twilight.

315 Cockchafer,
Melolontha melolontha

20-30 mm long. A widely distributed, plentiful and injurious insect. The grown beetle consumes the leaves of oak and other forest trees and the soil-dwelling larva feeds on roots, including those of crop-plants. Development lasts four years, and it is usual to speak of 'cockchafer years'. The beetles appear late in May and swarm in the evening. Larva, 389.

COLEOPTERA : Longhorn Beetles

316 *Prionus coriarius*

25-40 mm long and belongs, like the following 16 species, to the longicorns, so named because of the fancied resemblance of the long curved antennae to the horns of a goat. Occurs in decaying coniferous and broad-leaved trees, and is sometimes to be seen on herbaceous plants, such as fennel. Most often seen on tree trunks in southern England during July and August. The larvae live in the base of the trunk. Development takes 3-4 years.

317 Poplar Longicorn,
Saperda carcharias

18-28 mm long. Occurs about poplars and, more rarely, old willows, in the trunks of which the larva (391) lives. In Britain, its distribution is restricted, and it is probably commonest in the fens. The beetles are seen in July-September and swarm towards evening. Development takes 2-3 years.

318 *Cerambyx scopolii*

20-28 mm long. Uncommon. Most frequent near beech trees in which the larvae live. The beetles are abroad from the end of May onwards and feed in flowers.

319 Musk Beetle, *Aromia moschata*

23-32 mm long. Widespread but not common. From July to August may be seen on the leaves of willow trees or on flowers, especially those of umbelliferous plants. The larva is found in living willow trees.

320 *Stenocorus meridianus*

17-27 mm long. The male (320a) is more slender and often lighter in colour than the female. Widespread in broad-leaved woods but not common, and largely confined to southern England. Found in June-July on the flowers of umbelliferous plants, hawthorn, etc. The larva feeds in various trees.

321 *Saperda populnea*

10-14 mm long. The adult is common every other year as the larva takes two years to develop. Associated with aspen, more rarely with other poplars or willow. The feeding activities of the larvae inside the young twigs cause galls (600).

322 *Tetropium castaneum*

10-18 mm long. Occurs in coniferous woods, especially fir, from late in May till the beginning of July. The larva may be found both on living and on dead trees.

323 Timberman,
Acanthocinus aedilis

- 12-20 mm long. The antennae are very long, especially on the male (323a), where they are four times the length of

G

the body. The beetle occurs locally in Scotland and Ireland under the bark of fir, and is imported with pine timber. The larva feeds in dead coniferous wood.

324 *Callidium violaceum*
11-14 mm long. May be met with in dry, coniferous timber, either in woods or, frequently in saw-mills and new buildings. The beetle is about in June and August. Larva, 392.

325 **Wasp Beetle,** *Clytus arietis*
10-14 mm long. Resembles a wasp both in appearance and in its jerky movements. Widely distributed and very common in June on flowers and brush-wood in the sun. Breeds in various deciduous trees.

326 *Rhagium mordax*
12-22 mm long. Common in May-June on flowers in sunny parts of deciduous forests. Larva, 393.

327 *Strangalia quadrifasciata*
13-18 mm long. Common in woods and scrub during the summer where it may be seen on elder-blossom. Breeds in various broad-leaved trees.

328 *S. melanura*
7-11 mm long. Only locally common in England. Seen on flowers in June to July. The female is stouter than the male; and both sexes are entirely black except for the red-brown elytra.

329 *Alosterna tabacicolor*
6-9 mm long. Common on flowers, especially those of umbelliferous plants, in June-July in many woods. The larva occurs underneath the bark of various trees.

330 *Molorchus minor*
7-14 mm long. Easily recognised by its short elytra and bulbous femora. Common on the Continent in elder blossom during June, but very rare in Britain. The larvae live under the bark of fir-trees, and their old burrows are sometimes to be seen on fences.

331 *Pogonocherus hispidus*
5-6 mm long. Locally common on various deciduous trees from June until the autumn, and often seen also on ivy. Breeds in the thin branches of hard-woods.

332 *Phytoecia cylindrica*
8-12 mm long. Occurs in deciduous woodlands, where it is to be seen on various plants from June to August. It breeds in the stalks of umbellifers.

COLEOPTERA : Leaf Beetles

333 *Zeugophora subspinosa*
2.5-3.5 mm long. This, together with the following 14 species, belong to the leaf beetles which, as a rule, colonize leaves both when larvae and adults. *Z. subspinosa* is a woodland species which occurs on young aspen in the early and late summer. The larvae live communally in a hole eaten out of a leaf.

334 *Clytra quadripunctata*
7-11 mm long. Local but fairly widespread in coniferous plantations, and on such plants as willow, lime and aspen in deciduous woods. The larva (394) lives in a case and devours the grubs of ants.

335 *Chrysomela geminata*
6.5-7.5 mm long. Common on *Hypericum* in July-October. The larva lives free on the leaves. The beetles hibernate

and reappear in May.

336 *C. populi*
10-12 mm long and one of the larger leaf beetles. Common in May-August on poplars, and sallows where it occurs particularly on reddened leaves. Larva, 395.

337 *C. aenea*
6.5-8.5 mm long. Always uniformly coloured, but several varieties occur, including green, blue and bronze forms. Locally distributed on sallow. The beetles first appear in late summer, after which they hibernate and re-appear in spring.

338 *Phytodecta viminalis*
5.5-7 mm long. Very common on various kinds of willow in May-June. The larva lives on the underside of the leaf.

339 *Phyllodecta vulgatissima*
4-5 mm long. Common on various species of willow and sometimes appearing in such large numbers as to become a pest. The larvae eat into the undersides of leaves.

340 *Lochmaea capreae*
4-6 mm long. Occurs throughout the summer season on willow, sallow and birch, sometimes in such numbers as to be harmful in plantations. Widely distributed, but only locally common.

341 *Galerucella lineola*
4-5 mm long. Plentiful throughout the summer on willow; less frequent on alder and beech. Sometimes a pest in willow plantations.

342 **Alder Leaf Beetle,**
Agelastica alni
6-7 mm long. Of rare occurrence in southern England, on alder. The adults appear late in the summer, hibernate and resume their activities in the following spring. Larva, 396.

343 *Phyllobrotica quadrimaculata*
5-7 mm long. Very common around the common skullcap (*Scutellaria galericulata*), the leaves of which are eaten by the adults. They are about in June-July. The larvae live in the underground parts of the plants.

344 *Luperus longicornis*
4-5 mm long. Recognisable by the filiform antennae which, in the male, are longer than the body. Common in June on birch and alder, rarer on willow. The habits of the larva are not known with certainty, but it is believed to live on plant roots.

345 *Derocrepis rufipes*
2.3-3.8 mm long. Common in woodland glades and meadows containing the vetches and legumes upon which it feeds.

346 *Chalcoides fulvicornis*
2-2.5 mm long. A brilliant golden green beetle which is common on aspen in the summer. The insect hibernates in the adult state.

347 **Tortoise Beetle,**
Cassida rubiginosa
6-8 mm long. Very common all through the summer on such plants as thistles and burdock. Larva, 397.

COLEOPTERA : Weevils

348 *Platyrhinus resinosus*
9-12 mm long. Of rare occurrence in southern England, where it may be observed at any time in the summer, often sitting motionless on stumps or trunks of deciduous trees. Lives in old

fungi, especially those growing on ash.

349 *Strophosomus melanogrammus*
4-6.5 mm long. Very common on trees and bushes, especially hazel.

350 *Otiorhynchus singularis*
6-9 mm long. Common in May and June both on deciduous and on soft-wood trees. Sometimes appears as a pest in gardens on fruit trees or bushes.

351 *Cryptorhynchidius lapathi*
6-9 mm long. The legs are folded when the insect is at rest or alarmed. Very common in summer on willow and alder, more rarely on poplar and birch. Larva, 401.

352 *Dorytomus tortrix*
About 5 mm long. Widespread but local in summer on aspen, more rarely on other poplars. The larvae live in the catkins and pupate in the ground.

353 *Phyllobius calcaratus*
7-11 mm long. Where it occurs it is common on various young trees, especially alder. The adult eats holes in the newly sprouted leaves, and the larva lives in the soil and feeds on roots.

354 *Ph. argentatus*
5-7 mm long. Common in the spring on beech and hazel, in the leaves of which it eats holes (618), sometimes causing significant damage in young beech plantations.

355 **Pine Weevil,** *Pissodes pini*
7-9 mm long. Locally distributed in coniferous plantations. The beetles eat away the bark on young branches and lay their eggs in holes in the bark of old trees, often weakened or dead. The larvae (399) eat out passages between the bark and the wood.

356 **Spruce Weevil,** *Hylobius abietis*
10-13 mm long. Common, though local, on pine, where the beetles eat out holes in the bark of young trees (581) and may cause considerable damage. They sometimes make holes in the bark of deciduous trees, especially birch. The eggs are laid in the roots of pine stumps and the larvae (398) eat out passages between the bark and the wood.

357 **Nut Weevil,** *Curculio nucum*
5-8 mm long. Very common on hazel in June-July. The larva (400) lives in the nut. An allied species, *C. venosus*, is found in southern England on oak and its larva infests the acorn.

358 *Rhynchaenus fagi*
2-2.5 mm long. Common on beech, in the leaves of which it eats out characteristic small holes (616) immediately after they have opened. The larva lives inside the thickness of a beech leaf (617).

359 *R. quercus*
2.5-3.5 mm long. Like the foregoing but associated with oak (625). Other species are common on alder, birch, willow and hazel.

360 *Cionus scrophulariae*
4.5 mm long. Common on figwort (*Scrophularia sp.*); rarer on mullein (*Verbascum thapsus*). The larva is surrounded by a mucuous coat and lives on the outside of the leaves.

361 *Rhynchites betulae*
3-5 mm long. Common on alder and especially on young birch. The female rolls up part of a birch leaf (612) in which she lays from two to four eggs.

362 *Apoderus coryli*
6-7 mm long. Very common on hazel,

rarer on other broad-leaved trees. The female makes an elaborate leaf roll (611) wherein the egg or a pair of eggs is laid.

COLEOPTERA : Bark Beetles

363 Oak Bark Beetle,
Scolytus intricatus

2.5-3.8 mm long. Like the following seven species, *S. intricatus* breeds in or underneath the bark. It is associated mainly with oak although it also occurs on many other trees. Larval galleries, 402.

364 *Blastophagus piniperda*

3.5-4.8 mm long. Widespread but only locally common, chiefly on such conifers as pines (*Pinus*), silver firs (*Abies*) and spruces (*Picea*), and occasionally on larch (*Larix*). Breeds mostly in trees attacked by fungi. The beetle can also cause damage by eating holes into first-year apical growths which may die and break off. This may occur to such an extent that the tops of the trees become seriously deformed. Larval galleries, 408; apical damage, 586.

365 *Dendroctonus micans*

7-9 mm long and the largest of the bark beetles. Formerly uncommon, but in recent years it has become widespread in spruce plantations. On the Continent, where it breeds in living trees and has caused great damage. Larval galleries, 407.

366 Fir Bark Beetle
Hylurgops palliatus

2.5-3.3 mm long. Locally distributed in coniferous forests in spring and autumn, but causes little damage of significance as it both feeds and breeds only in dead trunks. Larval galleries, 404.

367 Spruce Bark Beetle,
Hylastes cunicularius

3.5-5 mm long. The species occurs locally in spruce plantations from May to July. Mainly young plants are attacked. The larvae feed in fresh stumps and roots. Larval galleries, 405.

368 The Printer, *Ips typographus*

4.2-6 mm long. Very common in pine, spruce and fir, but generally causes little damage as it tends to breed in diseased or felled trees. The male usually associates with two females. Larval galleries, 409.

369 *Pityogenes chalcographus*

2-2.3 mm long. Particularly common in spruce woods. By contrast with the foregoing it breeds chiefly in thin bark on young trees or in the tops of alder trees. Larval galleries, 406.

370 Ash Bark Beetle,
Leperesinus fraxini

2.5-3.2 mm long. Breeds in ash trees. The beetle eats out short passages in the bark on fresh branches but does not penetrate deep enough to cause appreciable damage. Larval galleries, 403.

COLEOPTERA : Beetle Larvae

371 Violet Ground Beetle,
Carabus violaceus

Up to 37 mm long. May be recognised by the reddish-coloured head. Met with especially in the autumn and as a rule is inactive through the winter. Like the adult beetle (233) the larva is typically carnivorous and hunts mostly at night.

372 *Cychrus scaraboides*, var. *rostratus*

Up to 20 mm long. Distinguished from all other ground beetle larvae by its shortness and width. Like the adult insect (234) lives mainly on snails. Met with especially in the autumn.

373 Tiger Beetle,
Cicindela campestris

About 20 mm long. Lives in a deep, vertical shaft in the ground. When in need of food, it climbs up this shaft until its head, with upwardly pointing mandibles, is level with the surface of the ground. By wedging its hind end, and a protuberance on its abdomen, against the wall of the shaft, it remains firmly fixed in position. It feeds by seizing small animals which pass over the hole and it probably takes two years to complete its development. Adult, 242.

374 Sexton Beetle,
Necrophorus humator

Up to 33 mm long. Lives on dead birds and small mammals which the adult beetles (254) have buried and near which they have laid their eggs. The larvae pupate in separate cells in the ground underneath the remains of the carcase.

375 Snail Beetle, *Phosphuga atrata*
16 mm long. Distinguished from other carrion beetle larvae by the long antenna. Probably lives exclusively on snails as does the adult beetle, which resembles the carrion beetle *Silpha carinata* (259) but has a longer head.

COLEOPTERA: Beetle Larvae

376 Glow-worm, *Lampyris noctiluca*
20-25 mm long. Like the adult insect (261), it is able to produce light, but only from a pair of circular luminous organs situated on the underside of the eighth joint of the body. Feeds exclusively on snails, the bodies of which

it liquifies by injecting digestive juices into the prey through hollow mandibles.

377 Soldier Beetle, *Cantharis fusca*
Up to 23 mm long. In the winter season is sometimes observed creeping under the snow, a habit which has given it the nickname of 'snow worm'. Omnivorous. Adult, 262.

378 *Chrysobothris affinis*
Up to 40 mm long and easily recognised by its characteristic shape. Lives in and on bark of oak trees, more rarely on beech. Adult, 270.

379 Wireworm, *Denticollis linearis*
About 20 mm long. Most of the skipjack wireworm larvae are very similar to each other recalling in their shape and movements those of the familiar meal-worm. Many species are found in woods under the bark, in stumps, in leaf-litter or below the ground. In most species, the larva probably lives three years. Adult skipjacks, 272-279.

380 *Tomoxia biguttata*
10-12 mm long. Lives in rotten wood infested by fungal mycelia, probably subsisting on these. The adult beetle (286) appears in June.

381 Cardinal Beetle,
Pyrochroa coccinea

Up to 35 mm long. The easily identifiable larvae live underneath the bark on dead trees or stumps, often communally. Adult, 280.

382 Raspberry Beetle,
Byturus urbanus

About 4 mm long. It is this larva which is the well-known 'worm' in raspberries. Adult, 287.

383 Seven-spot Ladybird,
Coccinella septempunctata
10-12 mm long. Becomes established on plants that are attacked by aphids, upon which it preys as does the adult beetle (297). The parti-coloured pupa (383a) is firmly attached to the underside of a branch, etc.

384 *Ernobius mollis*
6-7 mm long. Very common in dead branches of fir or pine, from which it bites away the xylem without damaging the bark (384a). Adult, 301.

385 Ant Beetle,
Thanasimus formicarius
10-15 mm long. A predator on bark beetles and their larvae, colonising dead coniferous wood. Adult, 266.

386 *Hylecoetus dermestoides*
Up to 25 mm long. A larva of unmistakable form, which eats out long passages in diseased or fallen tree-trunks. Adult, 304.

387 Lesser Stag Beetle,
Dorcus parallelopipedus
Up to 65 mm long. Usually lives in diseased beech stumps and in dead parts of old beeches. Adult, 311.

388 Rhinoceros Beetle,
Oryctes nasicornis
Up to 100 mm long. Not British. On the Continent, it feeds in heaps of rotting sawdust, wood shavings, etc., in saw-mills and, more rarely, in compost heaps. Like many large larvae which devour wood, it takes several years to develop. Adult, 314.

389 Cockchafer, *Melolontha melolontha*
Up to 50-60 mm long. A subterranean larva which destroys the roots and underground stalks both of woody and herbaceous plants, including crops. When three years old it pupates in the ground, usually in July. After 5-6 weeks the adult (315) emerges, but hibernates in the soil until the following spring.

390 Rose Chafer, *Cetonia aurata*
Up to 50 mm long. Distinguished from the other scarabeid larvae by its small head. Lives mostly in diseased stumps, but some allied species colonise ants' nests. Adult, 313.

391 Poplar Longicorn,
Saperda carcharias
Up to 45 mm long. During oviposition, the female beetle bites holes in the bark on a branch of poplar and lays an egg in each hole. The larva first eats out a cavity under the bark and then penetrates into the xylem where it makes long passages. The passages and the escape holes are oval in cross section. Adult, 317.

392 *Callidium violaceum*
Up to 22 mm long. Makes passages immediately under the bark on the dry wood of deciduous trees, subsequently eating its way about 5 cm into the xylem when it is about to pupate. Adult, 324.

393 *Rhagium mordax*
Up to 35 mm long. Eats out passages in the bark on stumps and dead trunks of various deciduous trees, more rarely in firs. The pupa (393a) lies in the hole underneath the bark which the larva has lined with a ring of wood shavings. Adult, 326.

394 *Clytra quadripunctata*
Up to 9 mm long. The larva lives in a case lined with its droppings. Lives in

the nest of the wood ant (436) whose larvae and pupae it devours. Related species, likewise enclosed in a sac, feed upon dead leaves. Adult, 334.

395 Poplar Leaf-beetle, *C. populi*

Up to 13 mm long. Lives free on poplar leaves, less often on those of willow. It may destroy everything except the veins. Adult, 336.

396 Alder Leaf-beetle,
Agelastica alni

Up to 12 mm long. Mode of life like the foregoing, but lives on alder, less often on hazel. Adult, 342.

397 Tortoise Beetle,
Cassida rubiginosa

Up to 9 mm long. Common on thistles. The larva has a forked process on its hind end which extends over its back. The droppings are lifted up in the fork and may form a large irregular lump covering the larva (397a). Adult, 347.

398 Spruce Weevil, *Hylobius abietis*

15-18 mm long. Eats out passages in the roots and under the bark of dead coniferous trees and stumps. Adult, 356.

COLEOPTERA: Larvae and Larval Galleries

399 Pine Weevil, *Pissodes pini*

Similiar to the foregoing but only half the size. Lives in the bark of fir trees. As the eggs are laid in a single mass and, as the newly hatched larvae radiate in all directions, the passages usually form a star pattern (as seen on the extreme right of the figure). Before pupating, the larva makes a chamber out of flakes of wood from which the grown beetle (355) later gnaws its way out.

400 Nut Weevil, *Curculio nucum*

8-10 mm long. The larva is the common 'worm' in hazel-nuts. When the kernel has been devoured, the larva bites its way through the shell and pupates in the ground. Adult, 357.

401 *Cryptorhynchidius lapathi*

The larva of this weevil (351) usually feeds in branches and young trunks of willow, more rarely in those of alder, birch or poplar. It first eats away the underside of the bark, producing brown spots (as at the top of the figure), and later eats its way into the pith.

COLEOPTERA: Bark Beetle Galleries

402 Oak Bark Beetle,
Scolytus intricatus

Mating takes place either outside the bark or in a special chamber eaten out by the male. The characteristic system of passages is initiated by either sex gnawing a hole in the bark. From this first passage the female bites out a row of recesses, in each of which she lays an egg, and the larva then eats its own passage which usually runs at right-angles to the first. Generally, the initial passage is 1-3 cm long and set horizontally to the length of the tree. Each larval passage ultimately terminates in a small widened chamber where pupation takes place. Such passages are found particularly in the branches and young trunks of oak. Adult, 363.

403 Ash Bark Beetle,
Leperesinus fraxini

Associated especially with ash. The main passage is double in the sense that it extends on both sides of the entrance hole. It runs at right-angles to the length of the tree and may be over 10 cm long, with larval passages of 2-5 cm in width. Adult, 370.

404 Fir Bark Beetle,
Hylurgops palliatus
Very common in coniferous trees, especially in spruce. The main passage is 2-7 cm long and this longitudinal passage has a short lateral branch close to the entry hole. The female remains in this side branch. As the eggs are not deposited equidistantly along the main passage, the larval tunnels present an untidy appearance. Adult, 366.

405 Spruce Bark Beetle,
Hylastes cunicularius
Common in the roots and stumps of spruce. The main passage runs length-wise for 5-6 cm and has a small chamber for the female. Adult, 367.

406 *Pityogenes chalcographus*
In Norway spruce (*Picea abies*), more rarely in other coniferous trees, the system of passages begins with a chamber in which the mating takes place. Since one male mates with several females, each of which excavates her own main passage, the system of tunnels becomes star shaped. The beetle (369) tends to make its passages where the bark is thin, so that these are found more especially on branches or on the trunks of young trees.

407 *Dendroctonus micans*
Occurs especially in spruce. Distinguished from other bark beetles by the fact that the larvae do not each form their own passages but make a large communal hole which gradually becomes filled up with excreta intermingled with resin and sloughed skins. Adult, 365.

408 *Blastophagus piniperda*
Common in fir. The main passage is longitudinal, often 8-10 cm long, and is provided with small air holes. The larval tunnels are very long and are regular near their origin but later become curved and irregular. Adult, 364, apical damage, 586.

409 The Printer, *Ips typographus*
Common in Norway spruce. The system of passages consists of longitudinal runs from 5-15 cm long which extent outward from a communal mating chamber (cf. 406). The larval galleries which are up to 5 cm long, are curbed and increase rapidly in width. Adult, 360.

HYMENOPTERA : Wood Wasps and Sawflies

410 Greater Horntail, *Sirex gigas*
12-40 mm long. The female, which may be recognised by her ovipositor, is larger than the male (410a). Often imported in foreign timber and now well established in Britain. Occurs during summer in coniferous woods. The female uses the ovipositor as an auger to bore holes through the bark a short way into the wood of felled or diseased fir or larch trees. She is said to lay a total of 100 eggs, each in a separate boring in the tree. The larva (438) eats out passages in the timber. Ichneumon parasite, 417, 445.

411 Lesser Horntail, *S. noctilio*
15-30 mm long. Similar to the foregoing but may also breed in Scots pine, which the larger species seems to avoid.

412 Birch Sawfly, *Cimbex femorata*
17-23 mm long. Common in May and
June in deciduous woods. The grown
wasps cut out deep cavities in the bark
on branches of birch, beech, ash, poplar
and other broad-leaved trees. Larva,
439.

413 Sawfly, *Lyda erythrocephala*
10-12 mm long. The female (413) has a
red head, the male (413a) blue or
yellow. Uncommon in fir woods during
April and May. Like other commoner
members of the same genus (e.g. *L. pyei*
and *L. nemoralis*) the larva (440) lives
with its fellows in a communal web on
the infested tree (583), within which
the individual insect may construct a
case of leaf fragments, like that of a
caddis-worm.

414 Pine Sawfly, *Lophyrus pini*
7-10 mm long. The male (414a) may
be recognised by the feather-like
antenna and by the absence of yellow
spots. Flies during May and June in fir
woods. The larva (441) feeds on Scots
pine.

415 Sawfly, *Rhogogaster viridis*
10-13 mm long. Very common in hardwood forests in June and July. The
larva lives both on woody and herbaceous plants.

416 Willow Sawfly, *Pteronus salicis*
7-10 mm long. Not rare in June-July
on willow and poplar, whereon the
larva (442) also lives. On willow there
are a considerable number of different
sawfly species, some of which cause
galls.

HYMENOPTERA : Ichneumons

417 Sabre Wasp, *Rhyssa persuasoria*
20-35 mm long excluding the ovipositor: with antennae and ovipositor
extended, the female attains a length
of 85 mm. The largest of our ichneumons, and may be found in coniferous
woods during summer. The female
parasitizes the deeply concealed larva
(438) of the Greater Horntail (410), in
some way sensing its presence and
inserting her slender, fragile-looking
ovipositor through the bark and wood
to reach it. As the penetration takes
time she runs a considerable risk of
being devoured by a bird. Larva, 445.

418 Parasite Wasp, *Ichneumon* sp.
About 10 mm long. There are
numerous ichneumon species in Britain,
most of them small and inconspicuous
even when adult. They may be more
in evidence when larvae, as a result of
their parasitic activities (447-449).
The species illustrated here lays its eggs
in the caterpillars of butterflies.

419 Yellow Ophion, *Ophion luteus*
15-20 mm long. Easily recognised by
the constricted curved hind end of the
body. It appears to parasitize the
caterpillars of noctuid moths.

**HYMENOPTERA: Burrowing
and Social Wasps**
420 Mournful Wasp,
Pemphredon lugubris
7-13 mm long. A burrowing wasp
which is common in woods from June
to August. It makes its nest in dry
stumps, with a preference for those of
decayed beech. The structure consists
of short side passages leading out of a
main gallery. It provisions it with live
aphids which it paralyses with its
sting, lays an egg and closes up the
nest. The larva devours the supply of
fresh meat and pupates without con-
structing a cocoon.

421 Digger Wasp, *Crabro vagus*
8-14 mm long. Lives like the foregoing but gathers flies as food for the larvae.

422 Spider-hunting Wasp,
Pompilus fuscus
7-13 mm long. Common on paths in woods from the early spring. Digs out its nest in the ground of the path and uses spiders as fodder for the larvae. The females hibernate in small holes in the ground.

423 Digger Wasp,
Anchistrocerus parietum
10-14 mm long. In many parts of the Continent, much in evidence from May to August. Related to the social wasps, which may be distinguished from the solitary digger wasps by their wings being folded lengthwise when at rest. The present species, however, does not form communities, but each female makes her own nesting places in holes and crevices. Food for the larvae consists of the small caterpillars of butterflies.

424 Red Wasp, *Vespa rufa*
Worker 12-14 mm long, drone 17-18 and queen 18-20 mm long. The red colour is never conspicuous, and affects mostly the base of the abdomen. Common in woods, where the nest is generally built underground, often in a cavity at the base of an old tree stump. As in the case of the other social wasps, the nest is started by a queen which has hibernated, and lasts for a single season only, as all the workers and drones die off in the autumn.

425 Tree Wasp, *V. sylvestris*
Similar in size to the next species and best recognised by the clear, yellow face (clypeus) which lacks the 'edging-tool' figure (426a). Not rare in woods and scrub. The nest is suspended from the branch of a shrub or tree, and is made of greyish wood-pulp.

426 Common Wasp, *V. vulgaris*
Worker 12-15 mm long, drone 16-17 and queen 17-19 mm long. Can be recognised by the pattern on the clypeus which takes the form of a black patch (426a) shaped like a gardener's edging-iron. Very common in woods, scrub and gardens. The nest is usually built underground or in a decayed tree stump. Scavenger, 476.

427 Hornet, *V. crabro*
Worker 22-24 mm long, drone 29-35 and queen 29-38 mm long. A denizen of woodlands containing decayed timber. Our largest species, recognizable by the brown hairy covering. Generally the nest is built in a hollow tree but sometimes under a roof or in a nesting box provided for birds. The wood-pulp of which it is made is taken from rotten wood and is very friable. As in the case of other social wasps, the larvae are fed on masticated insects, including flies, but the adults take sugar. When, in autumn, a shortage of insect food develops, the hornet larvae are killed and carried out of the nest.

HYMENOPTERA: Bees

428 Large Red-tailed Bumble-bee, *Bombus lapidarius*
Queen 20-27 mm long, drone 15-18 mm and workers of very different sizes. Common both in woods and elsewhere. The hibernated queens normally appear first in May and begin at once to make a nest in the ground. The workers appear during June, the new drones and queens in August. Immediately after mating in August the young queens hibernate inside stone walls and similar places. This,

and other bumble bees, may be infested by mites of the genus *Parasitus* (535).

429   Early Bumble-bee, *B. pratorum*
Smaller than the previous species, which it closely resembles. Found mostly in woods and gardens. The hibernated queens appear in April, the first workers in May, and new drones and queens in July.

430   Buff-tailed Bumble-bee,
    B. terrestris
The same size as *B. lapidarius*. Very common in woods and timbered gardens. The queens emerge from hibernation at the beginning of April, or, in a mild spring, as early as March.

431   Common Carder-bee,
    B. agrorum
The same size as *B. pratorum*. Variable in colour, but generally tawny and lacking the contrasting bands found in most species. It frequently looks unkempt. The name *agrorum* is misleading in that this is probably the most typical forest dweller among bumble-bees. Queens are about from the end of April. The nest is always on or above the ground, and the cells are enclosed in a covering of woven moss or grass.

432   Vestal Cuckoo-bee,
    Psithyrus vestalis
Similar in size and colour to *Bombus terrestris*, on which it is parasitic. May be recognised by the darker wings and more sluggish flight. The hibernated females come out in May and fly about looking for *terrestris* nests in which to lay their eggs. Another cuckoo-bee, *Ps. rupestris*, which resembles *Bombus lapidarius*, is parasitic on the latter.

433   Blue Osmia, *Osmia caerulescens*
8-10 mm long. A solitary bee, which

makes its nest in a hole in a tree stump, an earthen bank, or a mortar-course of a wall. Common from May to July and often seen on flowers. In Britain there are many kinds of solitary bees, some of which occur in woods and most of which are inconspicuous.

HYMENOPTERA: Ants

434   Red Ant, *Myrmica rubra*
Queen 4.5-7 mm, male about 5 mm and worker 3.5-5 mm long. Red ants have two nodes in the waist. Ubiquitous. The nest is found underneath stones or fallen branches or in tree stumps or hollow trees. The winged castes generally swarm in August. Mainly carnivorous, attacking and stinging other ants, capturing insects generally, but also keeping flocks of aphids from which honey-dew is 'milked'.

435   Orange Ant, *Lasius fuliginosus*
4-6 mm long. The nest is found in holes in trees, especially in willow, poplar and oak. The winged castes swarm on still, close days in June and July. After mating the males die, while the queens lose their wings and start a new community or return to the old one and supplement its egg-laying capacity. The community lives partly on honey-dew, secreted by aphids. Like the following ants this one has no sting but bites its victim and injects poison into the wound from its mouthparts.

436   Wood Ant, *Formica rufa*
Male and queen, 9-11 mm, worker 4-9 mm long. Frequent in coniferous woods, where it builds the well-known ant hill, of which the portion above ground is made up of pine needles. The superstructure may be up to 1.80 m high and 18 m in circumference, although it is usually smaller than this.

The underground section may extend a metre into the soil. A single ant hill may contain 100,000 individuals. Winged ants occur throughout much of the summer; they fly out on warm calm afternoons but do not form swarms. What are called 'ants' eggs', used as food for cage birds and aquarium fish, are dried pupae. The larvae and pupae are eaten by the grub of the beetle *Clytra quadripunctata* (394).

437 Hercules Ant,
Camponotus herculeanus
Females 14-17 mm, males 9-11 mm, and workers 5-12 mm long. Not common. The nest is made in a decaying tree stump and it may extend many metres into the trunk, causing additional weakening. Winged males and females occur in the nest throughout the year. Although they emerge from pupae in the summer, they do not leave the nest and swarm until the summer of the following year.

HYMENOPTERA: Larvae of Wood Wasps and Sawflies

438 Greater Horntail, *Sirex gigas*
Up to about 40 mm long. The larva eats out long galleries in fallen or decayed coniferous stumps and takes 2½-3 years to develop. Pupation often occurs in passages deep inside the trunks and the grown wasp (410) may have a long distance to gnaw through. It sometimes happens that the tree containing the larva or chrysalis is processed for building timber, with the result that the grown wasp emerges into an occupied room. The larval gallery and exit hole are circular in section. Parasitized by 417, 445.

439 Birch Sawfly, *Cimbex femorata*
Up to 40 mm long. Lives singly on various deciduous trees, particularly birch, the leaves of which it eats. The larvae of sawflies generally resemble moth and butterfly caterpillars, but the former have only two simple eyes and often six to eight pairs of prolegs, whereas the caterpillar of a Lepidopteran has twelve eyes and, at the most, five pairs of prolegs. The full grown larva spins a cocoon (439a) in which pupation takes place. Adult, 412.

440 Sawfly, *Lyda erythrocephala*
Up to 20 mm long. Distinguished from other sawfly larvae by the prominent antennae and the absence of prolegs. The larvae live communally in a web on fir trees, the needles of which they eat (583). Adult, 413.

HYMENOPTERA: Sawfly Larvae

441 Pine Sawfly, *Lophyrus pini*
Up to about 25 mm long. Found living in communities during the summer on Scots pine. Not only are the needles eaten, but the insect also attacks the new shoots which would otherwise develop into branches. It appears to be ignored by insectivorous birds and is a pest to the timber-grower. The full-size larva spins a cocoon (441a) either on the tree or, more often, on the forest floor; and the adult sawfly (414) emerges in the following June by making a lid in one end of the cocoon.

442 Willow Sawfly, *Pteronus salicis*
Up to 20 mm long. Common on willows, which often carry many larvae on a single leaf. If the leaf is disturbed the larvae simultaneously elevate their bodies. Adult, 416.

443 Pear Sawfly,
Eriocampoides limacina
About 10 mm long and resembling a

diminutive slug – hence the nickname 'slugworm'. It secretes mucus, which gives it a shiny look. Found on the upper surface of the leaves of various deciduous trees, most frequently cherry and pear. The leaves are skeletonised during feeding.

444 Oak Sawfly, *E. annulipes*

Like the foregoing, but the larva is covered with a light-coloured mucus and attaches itself to the underside of oak leaves, or more rarely, to the leaves or birch or willows.

HYMENOPTERA : Larvae of Ichneumon, etc.

445 Sabre Wasp, *Rhyssa persuasoria*

Up to about 25 mm long. Lives as an external parasite on the larva of the greater horntail (410, 438), by boring a hole in its skin and sucking the body fluids. It pupates in the gallery made by the host larva, after which the adult wasp (417) bites an escape hole through the side of the tree.

446 Larva of wasp parasitic upon weevil larva

The weevil larva is 13-14 mm long. Whereas the majority of parasitic wasp larvae live inside other larvae, especially those of butterflies, there are some that sit upon their prey and suck its juices from outside. The larva of *Rhyssa* (445) also behaves in this way.

447 Moth caterpillar containing pupae of parasitic wasp

The caterpillar is that of *Hyponomeuta euonymella* (70, 181). It is about 20 mm long and consists only of an empty skin enclosing numerous pupae of a parasitic wasp. The many larvae were produced in the first place by poly-embryony, i.e. each of the few eggs that were laid in the host divided to form several embryos. This phenomenon is not uncommon in such insects as aphids and ichneumons.

448 Aphid with cocoon of braconid, *Praon* sp.

Aphids are attacked by various parasitic wasps some of which may readily be noticed where there is a heavy infestation of aphids. In some cases distended aphids may be seen each with a round hole in the back. These are merely aphid skins, the living contents having been destroyed by the larvae of a braconid (*Aphidius* sp.). In other cases – as illustrated here – the dead aphid looks as if it were elevated on a tiny plinth, which is the cocoon of the braconid (*Praon*).

449 Parasitized caterpillar on spruce with pupae of braconid, *Microgaster* sp.

Many moth larvae are attacked by braconid wasps of the genus *Microgaster*. Frequently, the parasites are first noticed when they leave the body of a caterpillar they have killed to pupate on its outside, in characteristic small yellow cocoons.

DIPTERA : Gnat-like Flies

450 Fungus Gnat, *Sciara* sp.

A minute dark gnat, up to 4 mm long, which abounds in such places as accumulations of dead leaves or in moss. The larval stage (494) may colonise manure, plant roots, wild fungi, cultivated mushrooms, etc.

451 Crane-fly, *Tipula nebuculosa*

14-16 mm long. Flies in May-June. Many related species may all be recognised by the large wings and long legs. Most crane fly larvae (491) live in damp earth.

452 Crane-fly, *Nephrotoma crocata*
16-18 mm long and having shorter wings than No. 451. Common in June to August in damp woods, often in the vicinity of running water.

453 St. Mark's Fly, *Bibio marci*
11-13 mm long. The female (453a) can be recognised by her very large eyes. Swarms in sunshine in May at the edges and clearings in woods. The flight is neither rapid nor elegant, and the gnat looks as if its dangling legs were hindering the motion. The larva (492) lives in soil containing an abundance of decayed plant material.

454 Biting Gnat, *Aedes* sp.
The 'biting gnat' genus *Aedes* includes several species which are typical woodland forms. They live as larvae in spring pools and ditches from which the winged gnat emerges in May-June.

455 Fungus Gnat, *Mycetophilidae* sp.
4-8 mm long. A group of delicate, gnat-like flies involving over 60 genera and 450 species whose larvae (493) live in fungi, on rotten wood or in humus.

456 Long-horned Fungus Gnat,
Macrocera sp.
5-6 mm long. Recognisable by the enormously long, thin antennae which may be three times the length of the body. Not uncommon in summer on umbelliferous flowers. Larva, 495.

457 Winter Gnat,
Trichocera hiemalis
8-10 mm long. Common during most of the year in woods, where it is particularly noticeable when swarming on frostless winter days. The larvae

colonise the leaf-litter on the forest floor.

DIPTERA: Miscellaneous Flies

458 Horse-fly, *Tabanus bovinus*
18-23 mm long. One of our largest flies. Common in and near woods from May-August. The female sucks the blood from horses, cows and deer, and may also attack man. The male visits flowers for nectar. The larva is found in the ground and is carnivorous, feeding particularly on the larvae of beetles.

459 Cleg, *Hybomitra collina*
14-17 mm long. Common in the woods in June and the beginning of July, otherwise as above. The larger biting dipterons are often called gad flies.

460 Cleg, *Haematopota pluvialis*
8-13 mm long. Common both in woods and elsewhere in June -August and generally more plentiful in southern districts than in the north. The females seem to be particularly persistent in sultry weather, especially in thundery conditions.

461 Cleg, *Chrysops pictus*
8-11 mm long. Common close to and inside woods from the middle of June to the beginning of August, especially in proximity to water. The females are as bloodthirsty as those of the aforementioned clegs. The larva lives in water or mud.

462 Snipe-fly, *Rhagio scolopacea*
8-14 mm long. Very common the whole summer in woods where it may often be seen sitting on leaves and flowers or, head downwards, on trunks. It has the peculiar habit of suddenly taking to

flight and darting towards the observer. The larva (498) is carnivorous.

463 Robber-fly, *Laphria ephippium*
15-22 mm long. Widespread but uncommon. Seen especially in June and beginning of July in woods where as a rule it may be encountered sitting on a trunk. Lives on other insects which it catches in the air. A related species *L. marginata*, is comparatively common in deciduous woodland in southern England.

464 Robber-fly, *Neoitamus cyanurus*
12-17 mm long. Common throughout the summer in most parts of the woods where it is seen sitting on tree trunks, stumps, leaves, etc., waiting for insects flying by. The larva (501) lives in the forest floor.

465 Stilleto-fly, *Thereva nobilitata*
9-12 mm long. Common in summer on the margins of woods and along hedgerows, especially where there are stinging nettles. Probably feeds mainly on small flies. Larva, 500.

466 Greater Bee-fly, *Bombylius major*
7-12 mm long. Common in the spring on the edges of woods. A closely related species, *B. minor*, is not rare at the end of July and the beginning of August, both in woods and elsewhere. Both species often hover in the air while sucking the nectar from flowers, and *B. major* shows an apparent preference for those with a blue corolla. As larvae they are parasitic on solitary bees.

467 Empid, *Empis tesselata*
7-11 mm long. The largest British empid. It often occurs in considerable numbers from May to August on hawthorn blossom or umbelliferous flowers both in and near woods. It lives partly on nectar, and is partly a predator on other insects, including relatively large flies. The larva lives in the ground.

468 Long-headed Fly
Dolichopus claviger
5.5-6.5 mm long. Common in the summer on flowers and leaves in woods and scrub where it lies await for the small flies, aphids and the like, upon which it preys. The British fauna includes over 250 species of long-headed flies.

469 Wasp Fly, *Conops quadrifasciata*
11-12 mm long. Appears from about mid-June to August, and is often seen basking on the flowers of bramble. Easily recognised from its wasp-like appearance, its large head, and relatively long antennae. The larva is parasitic upon various wasps.

470 Hover-fly, *Rhingia campestris*
8-11 mm long. A common, odd-looking species, easily identified by its beak-like snout. Occurs throughout the summer on various flowers both in and out of woods. The eggs are laid in grasses and other low-growing plants overhanging cowdung, and the newly hatched larvae crawl into the dung and feed there. Belongs like the following eight species to the hover flies, of which nearly 250 species are known in Britain. The hover flies, drone flies or 'flower flies', make up a conspicuous part of our insect life during the summer.

471 Common Hover-fly,
Syrphus ribesii
9-13 mm long. Abundant from April to October in woodland glades and many other situations. Like most of the hover flies it becomes stationary in the air for a time after making a sudden short dart. On woodland paths hover-

flies may often be seen in a sunbeam. Larva, 503.

472   Hover-fly, *Helophilus pendulus*
10-14 mm long. Very common. Often seen on the inflorescences of ragwort and umbellifers; and the male has a habit of hovering over water, about a foot from the surface. The larva (505) resembles that of *Eristalis* and lives in the same way.

473   Hover-fly, *Chilosia albitarsis*
7-10 mm long. Abundant in spring in damp situations on such flowers as buttercup and marsh marigold. The larva lives in the stems of plants.

474   Hover-fly, *Myiatropa florea*
10-14 mm long. Very common from May to August especially on umbelliferous flowers in proximity to water. The larva is aquatic and has been recorded from pools of water standing in hollow beech-stumps.

475   Drone-fly, *Eristalis* sp.
9-16 mm long. The various drone flies are hover flies which superficially resemble bees and which may be seen on various flowers from the early spring until mid-autumn. The larva (505) is the well-known rat-tailed maggot.

476   Hover-fly, *Volucella pellucens*
12-16 mm long. Common in meadows and in the more open woods during June and July. The male is sometimes seen stationary in the air like the other hover flies, and the female observed taking nectar from such flowers as bramble and dog rose. The larva is found in the nests of social wasps, particularly *Vespa vulgaris* (426) where it performs the function of a scavenger, feeding on the excremental liquid of the wasp grubs, or the dead grubs themselves.

477   Hover-fly, *V. bombylans*
11-14 mm long. Like the foregoing, but the larva colonises the nests of bumble bees. Occurs in two forms outwardly resembling either *Bombus terrestris* (430) or *B. lapidarius* (428).

478   Hover-fly, *Xylota segnis*
10-13 mm long. Seen throughout the summer in woods and scrub where it runs about on the leaves and consumes the honeydew from aphids. When basking, with its wings folded over the abdomen, it could easily be mistaken for a wasp. Apparently not very well adapted for flying. The larva lives in rotten tree stumps, as is usually the case with other species in the genus.

479   Fruit-fly, *Ceriocera ceratocera*
8-9 mm long. Band flies, which take their name from dark bars on the wings, are small insects whose larvae live in plants where they sometimes occasion damage resembling galls. The larvae of the present species develop in the flower-heads of thistles: those of others eat into the leaves of various plants or live in stems or roots.

480   Marsh-fly, *Lyciella rorida*
4 mm long. A small fly with a metallic lustre, the larva of which lives in fallen hardwood timber.

481   Flesh-fly, *Pollenia rudis*
5-9 mm long. Very common until late in the autumn when it often invades houses situated close to woods. A sluggish fly which, as a rule, can easily be caught by hand. The larva lives as a parasite in earth-worms.

482   Muscid, *Mesembrina meridiana*
10-12 mm long. Common in and

around woods where it may be seen from early spring until late autumn, often sunning itself on a tree trunk or on the ground. It visits umbelliferous flower-heads for nectar. At each oviposition, the female lays a single large egg about 4½ mm long, containing a larva which hatches almost immediately. The egg is deposited in horse or cow manure.

483 Blow-fly, *Calliphora vomitoria*

9-12 mm long. Common in woods and gardens and indoors throughout the summer. The male is often seen on flowers or basking in the sun. The female lays her eggs on raw meat and carcases, and sometimes in wounds on living animals. The eggs hatch out in the course of a day. Larva, 504.

484 Flesh-fly, *Sarcophaga carnaria*

6-17 mm long, very variable in size. Abundant and ubiquitous in summer. The female is viviparous, depositing larvae instead of eggs on meat and carcases. The larvae disgorge a fluid which liquefies the meat; and this produces a medium in which the insects might drown but for the fact that their breathing-holes (spiracles) can be closed by fleshy lobes should they become submerged. Pupation occurs in the soil.

485 Parasite-fly,
Gymnochaeta viridis

7-12 mm long. A parasite fly which may be distinguished from the green-bottle (486) by the shape of its head. Common, especially in woods during the spring, where the males may often be seen basking, head downwards, on sunlit tree trunks. The maggot is parasitic on the larvae of certain weevils and noctuid moths.

486 Green-bottle, *Lucilia* sp.

There are several closely related species which are difficult to separate and which vary in size from 5-11 mm. Common throughout the summer both in and beyond woodlands, but are less frequently seen indoors than blow-flies (483). Many of them live similar lives to *Calliphora*, but the larva of one species is a parasite in toads. Another sometimes lays its eggs in the nostrils or ears of sleeping animals or man.

487 Muscid, *Hydrotaea irritans*

5-6 mm long. Very common indeed, and known to anyone who has walked in woods on a warm summer day. This is the fly that is attracted to human perspiration and swarms in numbers around one's head. Although it does not bite, it causes great irritation when it alights to sip the sweat. It is equally attracted to suppurations from sores or blood oozing from wounds or the mucous secretions from the nostrils and mouths of farm stock. The larva lives in manure.

488 Parasite-fly,
Eriothrix rufomaculatus

7-9 mm long. Perhaps the commonest of the Tachinid (parasite) flies and especially frequent in July and August on vegetation in woods along the coast but surprisingly little is known about its mode-of-life. The larva probably lives in that of the cockchafer (389).

489 Parasite-fly, *Dexia rustica*

7-12 mm long. Very common, from mid-May to August, on herbs and bushes along the brow of a wood or scrub. The larva is parasitic in that of the cockchafer. The eggs are laid on the ground, and the fly larvae quickly hatch and immediately disperse in search of a suitable host.

490 Bird Louse-fly,
Ornithomyia avicularia
4-5 mm long. Commonly found on a variety of birds, especially nestlings, including birds-of-prey, pigeons, jays and finches. When disturbed, it rapidly moves away with a peculiar sideways motion into the feathers of its host. Viviparous, the female giving birth to a single larva at a time, which soon pupates.

DIPTERA : Larvae of Gnat-like Flies

491 Crane-fly, *Tipula* sp.
Up to about 35 mm long. Leather-jackets – the larvae of crane-flies (daddy-long-legs) – are very common in the ground both in and out of woods. They feed partly on the dead remains of plants and partly on the roots and underground stems of living plants. Pupation takes place in the soil, and before the adult crane-fly (451) emerges, the pupa (491a) works its way towards the surface until it partly projects above ground.

492 St. Mark's Fly, *Bibio marci*
Up to about 20 mm long. Larval habit similar to the foregoing. Those species of *Bibio* whose larvae attack roots can become pests in gardens and farmland. Adult, 453.

493 Fungus Gnat,
Mycetophilidae sp.
White larvae up to 8-10 mm long, and known to all mushroom gatherers from the fact that they eat passages into the caps and stalks. The number of species is very large, and includes some which feed in fungi poisonous to man. Adult, 455.

494 Army Worm, *Sciara* sp.
For some unknown reason, numerous larvae of fungus gnats sometimes wander about in unison over the forest floor, setting up the so-called 'army worm', a moving column up to 6 m long. Normally the larvae live separately, many of them in decaying matter of some kind. Adult, 450.

495 Long-horned Fungus Gnat,
Macrocera sp.
6-7 mm long. This attenuated larva inhabits a kind of 'pipe' constructed out of loose webbing. Adult, 456.

496 Gall Gnat, *Mikiola fagi*
2-3 mm long. A dipteron which induces the formation of galls on beech leaves (622). Here a single gall is shown cut down the middle so as to expose the larva in the hollow space. Gall gnats are very small insects which are difficult to identify except from the vegetational features of the galls they cause.

497 Biting Midge, *Forcipomyia* sp.
4-5 mm long. These bristled larvae are common underneath bark, in sap flowing from wounds, rotting vegetation, etc. Often the bristles carry droplets of water.

DIPTERA : Miscellaneous Fly Larvae

498 Snipe-fly, *Rhagio* sp.
Up to about 20 mm long. Common on the forest floor where it hunts other larvae. Adult, 462.

499 Snipe-fly, *Xylophagus* sp.
Up to about 20 mm long. Recognisable from the long pointed head. Found underneath bark and in rotten wood where it hunts for other larvae. The adult, 10-15 mm long, is slender and black and rather like a parasitic wasp in appearance.

500 Stiletto-fly, *Thereva* sp.
Up to about 30 mm long with characteristic quick, snake-like movements. Common in the forest floor and in rotten wood, where it preys upon small invertebrates. It appears to have more segments than other insect larvae owing to the constriction of some of the abdominal segments in the middle. Adult, 465.

501 Robber-fly, *Neoitamus* sp.
16-18 mm long. Lives in the forest floor and hunts other insect larvae. Hibernates in the ground as a larva and pupates in the following spring. Adult, 464.

502 Muscid, *Fannia* sp.
5-6 mm long, easily recognised by its wide shape and the feather-like attachment. Species of the genus *Fannia* include the lesser horse-fly and latrine-fly and occur in many different places where animal and vegetable substances are rotting. Some species are not rare in the forest floor.

503 Hover-fly, *Syrphus* sp.
Up to 12-13 mm long. Looks and moves like a leech and is commonly met with on leaves attacked by aphids, upon which it feeds. The characteristic pupa (503a) is attached to leaves and twigs. Adult, 471.

504 Blow-fly, *Calliphora* sp.
Up to about 18 mm long. The well-known larvae (gentles) are very common in animal carcases. Under favourable conditions they can attain full size within a week of hatching, after which they pupate in the soil below the meat. The pupa (504a), as is usual among flies, is barrel-shaped. Adult, 483.

505 Rat-tailed Maggot of **Drone-fly,** *Eristalis* sp.
Up to 20 mm long, apart from the telescopic breathing-tube which may be several times as long as the body. Common in water polluted by organic matter, so that there is a low content of dissolved oxygen, and often found in hollow trees where water and decaying leaves collect. The breathing-tube enables the insect to supplement its oxygen supply from the atmosphere. Adult, 475 (see also 472).

ARANEAE : Spiders

506 *Hyptiotes paradoxus*
4-5 mm long. Very local (e.g. New Forest, Hants., England), where it generally occurs in box or yew and spins a triangular web (522) between lower branches which are often dead. Does not spin any sort of cover but sits on a branch that matches it closely.

507 *Anyphaena accentuata*
6-8 mm long. Abundant in woods on bushes and trees, and also on the forest floor. Does not spin any web but catches its prey on the run. The female spins a cocoon (523) for the eggs and sits near it until they hatch.

508 Jumping Spider,
Marpissa muscosa
9-10 mm long. The largest of our jumping spiders which as the name implies catches its prey by leaping on it. Occasional on trunks, fence posts and hop poles, where it stays under the loose bark. The body is flattened. Adult in April.

509 Zebra Spider, *Salticus scenicus*
6-7 mm long. Very common on sunlit tree-trunks, fences, walls, etc. Like 508 it spins a shelter in a crack or under-

neath bark, in which it remains when it is not hunting for prey.

510 Wolf Spider, *Lycosa amentata*
8-10 mm long. Abundant on parts of the forest floor exposed to sunshine, where it is especially noticeable in the early spring. One of the hunting spiders which surprise their prey by a rapid swoop. They do not spin a snare but, like all spiders (even those living on the ground) provide themselves with a spun thread as protection against falls. The female spins a globular web in which she lays her eggs and which she carries with her, attached to her hind legs.

511 Hunting Spider,
Pisaura mirabilis
11-13 mm long. A hunting spider which is common in open parts of woods where there is rich vegetation. The female spins a very large egg pill (511a) which she carries by means of her poison fangs and not attached to her hind legs as in 510. In addition, she spins a bell shaped structure about 5 cm high, among the upper parts of grasses and other herbs wherein she remains with the egg pill. When the young spiders hatch they spin threads radiating from the bell, thus forming a communal web. After about a week, the young spiders disperse.

512 *Micrommata virescens*
11-13 mm long. Common in oak woods and in scrub, where it supports itself on the leaves of low bushes and herbs. A quick-moving spider which leaps on to its prey. Although the male (512a) is more conspicuous than the female, he is more rarely seen. When she is ready to lay her eggs the female withdraws to a shelter formed by spinning several leaves together.

513 Crab Spider, *Misumena vatia*
Female 10-12 mm long, and the male (513a) considerably smaller. Common on bushes throughout southern Britain. It mainly attacks insects visiting flowers. The egg cocoon is held in a folded leaf which the female tightly secures by silk.

514 *Diaea dorsata*
Female 7-8 mm long, the male smaller. Very common on leaves of box and other broad-leaved shrubs in many districts south of Yorkshire, but difficult to observe. Resembles 513 in its general habits.

515 *Meta segmentata*
8-9 mm long. One of the commonest spiders. Its characteristic snare (524) is to be seen everywhere, in low vegetation, between the branches of bushes and trees, and on the window-frames of sheds and houses. The spider sits close to the snare, usually with its first pair of legs extended forward and its second pair backwards. The eggs are deposited in a spherical cocoon beneath stones, fallen leaves, or in thick moss.

516 *Zygiella atrica*
8-10 mm long. Very common in woods and gardens as well as along hedges, where the characteristic snare (525) may be seen hanging in trees and bushes. The spiders mature in autumn, and, after mating, the female lays her eggs in a vault-shaped web underneath bark or in crevices, or among the needles on fir trees. The young spiders appear in the following spring.

517 *Araneus umbraticus*
12-13 mm long. Darker and flatter than the cross spider (518). Abundant. It hides itself during the daytime in

narrow crevices in the bark of old trees. The snare is very characteristic in being markedly asymmetrical, the radii converging near to the crevice in which the spider sits. Here it over-winters, and the eggs are laid in the following year.

518 Cross Spider, *A. diadematus*

12-15 mm long. Very common in woods, hedgerows and gardens, its large web (527) being familiar to everyone. Pairing takes place in August and September, and the female lays her eggs in a vault-shaped cocoon underneath bark or in a crevice. Often the web is covered with small leaf particles, or the cocoon itself is covered by a large white web on which the female stays until she dies later in the autumn. The eggs remain over winter and hatch in the following year.

519 *A. cucurbitinus*

7-8 mm long. Abundant throughout Britain on bushes along the edges of woods and hedges, especially on the south side. The snare is quite small, often not larger than a leaf, and easily overlooked. The egg cocoon is more conspicuous (528).

520 *Cyclosa conica*

7-8 mm long and easily recognised by the characteristic shape of the abdomen. Very common both in broad-leaved and coniferous woods. Its snare is easy to recognise by reason of the fact that a vertical white band (the *stabilimentum*) is spun through part of it. Often the remains of meals are left along the same vertical strip. The eggs are deposited during July and August in a small spherical cocoon attached to a dead twig.

521 *Linyphia triangularis*

6-8 mm long. Widespread and abundant, and especially plentiful in pine woods where the characteristic web (526) is everywhere to be seen. The egg cocoon is deposited in the autumn on the forest floor, where it is often attached to fallen pine-needles.

ARANEAE : Spiders' Webs

522 *Hyptiotes paradoxus*, snare

This snare, attached at only three points usually to dead pine branches, is impossible to confuse with any other web. Spider, 506.

523 *Anyphaena accentuata*, egg cocoon

Generally found on the underside of a leaf, the edge of which is bent over. The female sits with the cocoon until the eggs are hatched. Spider, 507.

524 *Meta segmentata*, snare

The snare is characterised by the circular hole in the centre. After the spider (515) has spun the web it bites away the radial threads from this area.

525 *Zygiella atrica*, snare

This snare cannot be confused with any other. It is unique in having the circular sticky threads omitted from a radial sector near the top. In the middle of this sector runs a signalling thread from the centre of the web to the shelter where the spider is lurking. Vibrations set up by a snared insect are transmitted to the spider along it. The egg cocoon (525a) is commonly spun over with tight net of white silk. Spider, 516.

526 *Linyphia triangularis*, snare

The snare consists of a horizontal sheet having barrier threads above and supporting threads below. The barrier threads cause flying insects to fall on the sheet, which is not sticky, and the spider runs to seize its prey before the latter recovers. As the barrier mesh-work also hampers the spider's own movements, it travels along the under-side of the web between the supporting threads (which are further apart than those of the barrier) and attacks its victim through the web. Spider, 521.

527 Cross Spider,
Araneus diadematus, snare

As is generally the case with orb webs, the snare is spun from two kinds of thread. Those which are not sticky form the peripheral threads, the radii and the middle part of the web where the spider sits; those provided with sticky globules are the snare threads which form a close spiral over most of the web. Spider, 518.

528 *A. cucurbitinus*, egg cocoon

This is commonly seen in the summer and autumn along forest paths and at the edges of woods. It is placed at the tops of various grasses which are bent inwards towards the cocoon and held in position by threads. Spider, 519.

OPILIONES : Harvest-Spiders

529 Harvestman, *Liobunum rotundum*

Female 5-6 mm and male 3-4 mm long. Very common both in and out of woods, to be found not only on the ground but also higher up in the vegetation. They mature in the late summer. The eggs are laid in a heap on the ground, or in moss, and the young appear in the following spring.

530 Harvestman,
Lacinius ephippiatus

Female 6 mm and male 4 mm long. Habits are similar to the foregoing, but they keep more to the forest floor.

531 Harvestman,
Nemastoma lugubre

2-2.5 mm long. Common in woods, where it lives in moss or under withered hardwood leaves or branches, etc., on the forest floor. Mature individuals can be found almost the whole of the year and young ones late in the summer.

CHELONETHI: Pseudo-scorpions

532 Pseudo-scorpion *Neobisium muscorum*

3 mm long. Inconspicuous, but actually common in many woods where it occurs in leaf litter on the forest floor, in moss cushions, and sometimes in old nests. Like true scorpions it has pincer-like chelae, but it is without the poison vessel at the point of the abdomen. It largely feeds on mites.

ACARI: Mites and Ticks

533 Sheep Tick, *Ixodes ricinus*

1-2 mm long when not gorged with blood, the female expanding to 10 mm after feeding (533a). Common in woods with a thick herb layer. The life-history is typical of closely related forms, where the young acarines sit on plants until an animal passes which they can colonise. They attach them-selves to this and creep over the body, until they find a sheltered place where the skin is thin. On a dog this may be behind the ears: on man in the fork of the legs or in the arm pits. After about a week, a female sheep tick has absorbed so much blood that she is ready to lay eggs and she then drops off the host.

534 Red Earth Mite,
Trombidium holocericeum

Up to 4 mm long. Very common in deciduous forests early in the spring. It is predatory on mites, collembolans, etc. As in the case of other mites, the larva has only six legs. It is plentiful on grass late in the summer, and sometimes bores into the skin of mammals, including man, on which it causes an irritating rash. The larva is nicknamed the 'harvest mite'.

535 Mite, *Parasitus* sp.
About 1 mm long. These mites are peculiar in that, as larvae, they travel from place to place on the bodies of various flying insects. One species, for instance, is commonly seen on the beetle *Geotrupes stercorarius* (312) others on bumble bees (428-431). The grown mites live in the ground, in the nests of bees, etc. Their feeding habits are obscure and require further study.

536 Mite, *Banksia tegeocrana*
Barely 1 mm long. The young animal (536a) differs greatly from the adult. *Banksia* belongs to the 'armoured' mites which may be found in large numbers in leaf litter and the upper part of the soil. Woodland soil may contain as many as 10,000 per square metre.

537 Mite, *Camisia palustris*
About 1 mm long. Like other 'armoured' mites, this species has a protective covering of a putty-like substance.

538 Mite, *Galumna climata*
A little over 0.5 mm long and easily recognised by the large paired projections directed forward from the abdomen. Many 'armoured' mites have bizarre shapes and the shield may be embossed with small mouldings of the putty-like material forming a fine pattern.

539 Mite, *Phthiracarus* sp.
About 1.5 mm long and characterised by the fact that, when the legs are folded, the head shield covers them completely. When in this position, the mite looks something like a diminutive frog.

540 Oribatid Mite,
Oribata geniculata

1-1.5 mm long. The young oribatid (540a) covers its back with a thick layer of dirt. Oribatid mites are essentially herbivores, feeding upon plant refuse and fungal hyphae. They are of significance in transforming organic matter which accumulates on the forest floor.

MOLLUSCA : Slugs

541 *Limax marginatus*
About 60 mm long. Colour generally grey-green. Like the next two species can be distinguished from slugs of the genus *Arion* by the fact that the breathing hole is situated behind the mid-line of the mantle. Common in beech woods where large numbers may be seen in the summer and autumn on trunks when the weather is wet. In dry weather they hide in crevices and under bark.

542 *L. tenellus*
Up to 50 mm long. Common in the autumn (October-November) on the floor both of deciduous and of coniferous woods, but the young snails lead a concealed existence in leaf-litter or a little below the soil surface so that they are not often seen. The eggs are laid late in the year, over-winter and then hatch in the following spring.

543 *L. maximus*
Over 100 mm long. Very common in woodlands and elsewhere but, as a rule, appears only in wet weather or at night. The clear slimy-looking eggs are laid in a heap within a cavity in damp ground, or under branches or bark. The eggs over-winter but the adults die off in cold weather.

544 Red Slug, *Arion rufus*
Up to 150 mm long. Widely distributed in Britain often near gardens; and in parts of Europe has evidently increased its range through human agency. Usually reddish or yellowish, but a black form occurs which is difficult to distinguish from 545 except by dissection.

545 Black Slug, *A. ater*
Up to 130 mm long. Usually intensely black, but may be white, or white with a black back, more rarely grey, brown or reddish. Very common in woods where, in damp weather, it may be present in considerable numbers on roads and paths. Omnivorous. The eggs, oval in shape, measure 4 x 5 mm, have opaque shells, and are laid on the ground in clusters of 20-50. Eggs, or young slugs, survive the winter, the adults being killed off by hard frost.

546 Brown Slug, *A. subfuscus*
Up to 60 mm long. The slug is actually grey, but the yellowish slime it exudes makes it appear brown. Widely distributed especially in deciduous woodland, where it mostly stays on the ground but in wet weather may be seen crawling up the trunks, especially those of beech and ash. Life history similar to that of 545.

547 *A. circumscriptus*
30-50 mm long. Common and widely distributed under stones, fallen branches, leaves, etc., both in woods and beyond. Probably the only British slug which normally hibernates as an adult.

MOLLUSCA : Snails

548 *Carychium tridentatum*
Shell scarcely 2 mm high. Common under moss and leaves on the surface of the ground in woods. Related to some of the aquatic and marshland snails and, like these, has only one pair of tentacles with eyes at their base.

549 *Columella edentula*
Shell 2.5-3 mm high. Common among moss and leaf-litter in undisturbed ground on the forest floor. Sometimes seen in large numbers creeping up beech and ash trunks.

550 Chrysalis Snail, *Vertigo* sp.
Shell 1.5-2 mm high. Several species of these small snails occur in Britain, and most of them can be found on the forest floor underneath branches, stones, dead leaves, etc. All have the mouth of the shell indented with four to nine teeth.

551 *Cochlicopa lubrica*
Shell 7 mm high. Common on damp and moist ground in woods and elsewhere. Often to be seen during autumn in large numbers low down on walls, on large stones or similar places, especially where there is a thick growth of grass. A smaller species is found on drier ground.

552 *Punctum pygmaeum*
Shell 1.5 mm wide. The Latin name of this, the smallest British land snail, is apt, as it is only about the size of a pinhead. Common on the forest floor in wet leaf-litter, where it is easily overlooked.

553 *Discus rotundatus*

Shell 7-8 mm wide. Easily recognised by the flat, tightly rolled shell with alternating lighter and darker spots. One of the commonest snails in woodland débris.

554 **Crystal Snail,** *Vitrea crystallina*

Shell 3 mm wide. The animal, which is white with a blue-grey head, can distinctly be seen through the clear, glassy shell. Common on soil rich in organic matter, where it occurs particularly in the top layer of mould and also on the underside of rotted branches.

555 **Glass Snail,** *Vitrina pellucida*

Shell 6 mm wide, and almost as clear as glass. Widespread in moss, woodland débris, etc. In the summer only young snails are about, and the species does not become noticeable until autumn, after which it may be seen during most of the winter whenever the weather is dry and mild.

556 *Euconulus fulvus*

Shell 2.5 mm wide. A brown, glossy shell shaped something like a spinning-top. Common on the forest floor underneath withered leaves, fallen branches, etc. May be observed early in the winter if there is no frost.

557 *Retinella nitidula*

Shell 8 mm wide. Almost opaque and with only a slight gloss. Very common on damp litter on the forest floor. A partly carnivorous species eating worms and other snails in addition to fungi and green plants.

558 *Oxychilus allvarium*

Shell 5-7 mm wide. Translucent, orange-coloured and glossy. Not as common as 557, but of similar habits.

Locally frequent in deciduous woods. Sometimes smells of garlic (hence *alliarium*).

559 *Perpolita hammonis*

Shell 3.5-4 mm wide and easily recognised by the fine radial ribs on the top side. Occurs both in deciduous and coniferous woods underneath fallen branches, dead leaves, etc.

560 *Retinella pura*

Shell 4-5 mm wide, and can vary in colour from dull white to brown. Common in deciduous woods where it occurs in débris on the forest floor.

561 *Ena obscura*

Shell 8-10 mm high. Very common in hardwood forests where it may often be seen on such trunks as beech and ash, and also on the forest floor under fallen branches. Young snails nearly always have the short, cone-shaped shell covered with earth.

562 **Little Coiled Snail,**
Clausilia bidentata

Shell 9-10 mm long, with fine stripes and the mouth sinistral (on left). Common in most kinds of woodland, particularly on the bark of deciduous trees.

563 **Common Coiled Snail,**
C. pumila

Shell 12-13 mm high, with prominent ribs. Can be found both on the forest floor and on tree trunks, especially those of beech and ash. It is commonly in summer and autumn that the snails are seen climbing the trunks.

564 **Large Coiled Snail,**
Iphigena ventricosa

Shell 15.5-18.5 mm high, with very rough ribs. Lives in damper places than

Clausilia and less often found on tree trunks.

565 Smooth Coiled Snail,
Cochlodina laminata

Smooth shell 14-18 mm high. Very common in many deciduous woods, where it may be seen in particularly large numbers on beech trunks. It hibernates in the forest floor.

566 Hairy Garden Snail,
Trichia hispida

Shell 7-9.5 mm wide, covered with fine hairs. The full grown individuals may be recognised by the mouth of the shell being thickened with a white rim. Common both in and out of woods, especially at places where there is a luxuriant undergrowth.

567 Hazel Snail, *Perforatella incarnata*
Shell 12-15 mm wide, and recognisable by almost closed 'navel' (the hole to be found on the underside of many shells). Occurs in deciduous woods on damp, organically rich ground, and tends to remain concealed.

568 *Helicogonia lapicida*
Shell 15-19 mm wide. A brown, strongly keeled shell with a white, reflexed lip. Widespread in damp woods, where it may be seen in summer and autumn on the trunks of trees. The winter is spent in a state of torpor in the forest floor, the mouth of the shell then being closed by a mucous epiphragm.

569 Bush Snail, *Eulota fruticum*
Shell 15-21 mm wide. The shell is off-white or brownish, with an open navel, and so transparent that the dark spots on the animal's body can be seen through it. Widespread and locally common in warm damp places in woods and scrub, especially where there is a rich herbaceous flora.

570 *Arianta arbustorum*
Shell 15-24 mm wide. Recognised by the white, reflexed lip and spiral ridges crossed by striae. Widespread through-out Britain in such damp places as woods on heavy soil, with a dense growth of herbaceous plants. Particularly common in stinging-nettle beds.

571 White-lipped Hedge-snail,
Cepaea hortensis

Shell 14-23 mm wide. Very variable in colour and marking. The ground colour may be white, yellow, or reddish usually with five or less dark bands which may be fused together. The lip is white and reflexed. Very common throughout the country, in woods, hedges, scrub and open land. Often seen low down on trunks.

572 Black-lipped Hedge-snail,
C. nemoralis

Shell 16-17 mm wide. Ground colour and banding similar to *hortensis*, but lip usually dark. General habits also similar, but probably not quite as common. Late in autumn both species hibernate, digging themselves a little into the ground, and closing the mouth of the shell by a mucous epiphragm strengthened with limestone.

573 Roman Snail, *Helix pomatia*
Shell 36-49 mm wide. The largest British land snail. Locally abundant on soils with a high calcium content, often in the overgrown margins of woodlands close to old houses. Hibernates like *Cepaea* just below the soil surface, but the epiphragm contains so much lime that it becomes a hard lid (573a). After being discarded, the epiphragm may persist for several months on the ground.

BOTANICAL ILLUSTRATIONS

Illustrations Nos. 574-686 give examples of how invertebrate animals affect woodland plants. As a rule only the results are indicated, not the animals causing them, since most of the causers are either small or have little to characterise their outward appearance. In cases where the animals are relatively conspicuous they are shown in the preceding plates and appropriate cross-references given in the text.

FERN

574 Intwined point on fern frond
A common phenomenon due to the larva of the dipteron, *Chirosia parvicornis*, 3-4 mm long when adult, which is related to the house fly. The larva occupies a position in a hole near the point of the frond, thereby hindering its normal expansion.

CONIFERS

575 Gall on the branch of spruce
On young shoots of spruce, often sitka spruce, galls 5-6 cm long may be found which are caused by the aphid *Gilletteella cooleyi*. The galls are often bent. Some generations of the aphid migrate to Douglas fir where they do not cause galls.

576 Small pineapple gall on spruce. At the point of shoots a yellowish white gall about 10 mm long may often be seen which later becomes brown and ligneous (576a). This is produced by the aphid *Cnaphalodes* sp. which is heteroecious with larch, on which it does not cause similar swellings, but has the effect of distorting the needles.

577 Pseudocone gall on spruce
Up to 35 mm long. A gall which superficially resembles a cone or pineapple. Composed, like 575 and 576, by a basic part of the needles swelling up. Due to the aphid *Adelges abietis* which also lives on larch (46) without causing galls.

578 Deformed needles
May often be seen as deformed and irregular needles on some of the shoots of silver and Norwegian fir. This is due to their being sucked by the aphid *Mindarus abietinus*, which is abroad only in May and June. During the remainder of the year it remains as an egg in the fir shoots.

579 Brown needles
White or brown needles are often seen on the shoots of red fir, held in position by a web. This is due to the larva of the microlepidopteran *Eucosma tedella* which hollows out the needles, causing them to lose their green colour.

580 Bent top shoots and attacked cones
Young fir trees are sometimes seen to have their top shoots bent down. This

is due to the larva of a moth *Dioryctria abietella* which has hollowed them out. The droppings are expelled and may be seen hanging loosely outside the shoot. Also attacks the cones. Adult, 79.

581 Bark stripped off conifers
The characteristic gnawing by the spruce weevil, *Hylobius abietis* (356) may often be seen on the bark of young trees.

582 Whitened needles
When some of the needles on an otherwise fresh shoot of fir become yellow and later brown, finally falling off, this is usually due to their being attacked by the larva of a gall-gnat *Thecodiplosis brachyntera*. The needles are swollen at the base (582a).

583 Matted needles
Often due to the larvae of sawflies, for instance *Lyda erythrocephala* (440). For the adult wasp, see 413. Other species of *Lyda* may also be encountered.

584 Resin 'galls'
Clusters the size of nuts may often be seen on the young shoots of fir trees. These consist of a web impregnated with resin. Underneath it is a deep groove on the twig which contains the larva of the microlepidopteran, *Evertia resinella*. The structure is not a gall in the true sense, as it does not involve proliferation of plant tissues.

585 Drooping top shoot
Often due to the fact that the shoot has been hollowed out by the larva of the moth *Evertia buoliana* (81), thereby killing it. The insect especially attacks young trees and can cause considerable damage.

586 Red needles
Shoots with conspicuous red needles may often be seen on mountain pines. On examining such a shoot more closely, a little resin may often be noticed exuding from the twig. Here the beetle *Blastophagus piniperda* (364) has bored its way in and begun to eat the pith in the shoot, which becomes hollowed out and dies.

587 Green gall
On the young branches of Scotch pine, irregular gall-shaped swellings may often be seen. These are due to a gall mite *Eriophyes pini* having bored its way into the bark.

588 Juniper 'berries'
The juniper 'berry', about 10 mm long, is not a fruit, but a gall induced by the gall-gnat *Oligotrophus juniperinus*. It consists of a cluster of needles enclosing the larva, and has sometimes been used as a remedy for whooping cough.

WILLOW

589 Curled edges of leaf
Due to the sawfly *Pontania leucaspis*, the larva of which lives inside the curled edge. Very common on various species of willow.

590 Elongated leaf gall
Galls with thin walls up to 20 mm long, equally apparent on both side of the leaf. Due to the sawfly *P. vesicator*. See also 598.

591 Hairy leaf gall on under surface
Due to the sawfly *P. pedunculi*. A similar leaf gall with longer hairs and often with red markings is attributable to another sawfly, *P. joergenseni*.

592 Spherical leaf gall
A round ball of about 10 mm size. Due to the sawfly *P. viminalis*. Common on several different species of willow.

593 Globose gall on branch.
Clearly delimited galls up to 15 mm long on the young shoots, enclosing many small pockets of larvae. Caused by the gall-gnat *Rhabdophaga salicis*.

594 Camellia gall
The leaves at the point of the willow shoot are very tight together and form a compact rosette. Caused by the gall-gnat *Rh. rosaria*, the larva of which develops inside the shoot tip.

595 Rolled leaf margin
The inturned edge is often of variegated yellow and red colour. Due to various gall-gnats of the genus *Dasyneura. sp.*

596 Pouch gall on lamina
These are often arranged in rows, each with a little red opening on the part projecting from the underside of the leaf (596a). Caused by the gall-gnat *Iteomyia capreae*.

597 Small hairy leaf gall on lamina
Galls of 1-2 mm in size on the upper surface of the leaf. Due to a gall-mite *Eriophyes tetanothrix*.

598 Bean gall on willow leaf
These may resemble the gall 590 but are only 10 mm long and have thick walls. They are green at first and later red. Caused by the sawfly *Pontania capreae*. One of the commonest kind of galls on willow.

599 Gherkin gall on stem
Galls on twigs of various species of willow often resembling gherkins irregularly bent. Met with especially on bay-leaved willow (*Salix pentandra*), where it may occur in large numbers on a single tree. Each gall contains up to 5 larvae of the hymenopteran *Euura pentandrae*. As the galled branches usually die, it may cause considerable damage to the infected tree.

POPLAR

600 Branch gall
Twigs and thin branches of poplar, less often those of willow often have a swelling in the shape of a spindle about 25 mm long which is due to the larva of the beetle *Saperda populnea* (321) which feeds inside the twig. There may be several of these galls on one branch.

601 Midrib gall on poplar leaf
This gall, 20 mm long and 10 mm high, occurs on the upper side of the leaf while, on the underside, a groove develops in the midrib. It is produced by the aphid *Pemphigus filaginis*, the spring generation of which lives on poplar and the summer brood on various compositae.

602 Spiral gall on petiole
Up to about 30 mm long. An unmistakable gall on the petiole caused by the aphid *P. spirothecae*.

603 Purse gall on petiole
Up to 15 mm long. Usually on the petiole, a pouch-like gall with an opening at one end, induced by the aphid *P. bursarius*.

604 Vein gall on lamina
Due to the gall-gnats of the genus *Harmandia*. Galls with an opening on the upper side are attributable to *H. cavernosa*.

605 Vein gall on lamina

Compare with 604. Galls which are generally spherical on the upper surface, with a corresponding recess on the underside (605a), are due to the gall-gnat *H. globuli*.

606 Spherical petiolar gall

An almost spherical swelling of the leaf stalk is due to the gall-gnat *Syndiplosis petioli*. Not to be confused with the two aphid galls, 602 and 603.

607 Swollen leaf glands

The two glands at the base of the leaf become red and swollen when colonised by the gall-mite *Eriophyes diversipunctatus*.

608 Hairy patches on leaf

Spots of felted hairs on the underside of the leaf, at first red and later brown in colour, corresponding to arched portions on the other side, are due to the activities of the gall mite *E. varius*.

ALDER

609 Small leaf gall

Numerous small spherical galls, at first yellow and later red, often occur on the upper side of alder leaves, and are caused by the gall-mite *E. laevis*.

610 Hairy spots on leaf

Hairy patches sometimes develop on the underside of alder leaves, at first whitish-yellow and later a rusty brown. They are caused by the activities of the gall-mite *E. brevitarsus*, and contain acarines in various stages of growth.

611 Rolled leaf

Rolls may frequently be seen on hazel, each formed from a single leaf. These are caused by the weevil *Apoderus coryli*, whose larvae develop inside the roll. Adult, 362.

BIRCH

612 Roll-gall

A characteristic cornet-shaped leaf roll on birch and, less frequently on alder, is induced by the weevil *Rhynchites betulae* (361). The roll is initially green and later turns brown.

613 Leaf-miner

Extensive, light coloured patches in a leaf are due to a moth larva *Eriocrania sparmanella*, having burrowed under the epidermis and into the mesophyll. The excretion of the larva is seen as arched stripes spread through the patches. Very common.

614 Leaf-miner

The mine is seen as a sinuous passage with dark excreta down the middle. It is produced by the feeding activities of the caterpillar of a moth *Lyonetia clerckella*, and is very common.

615 Leaf-miner

Sometimes there is seen on birch leaves a gallery which begins as a narrow channel and later on broadens considerably. This is due to the larva of the tiny leaf-mining fly *Agromyza alnibetulae*. The droppings accumulate along the margin of the gallery owing to the fact that the larva lies on its side when feeding whereas the moth caterpillar in 614 lies on its ventral surface.

BEECH

616 Small holes in leaf

Even immediately after expanding, many beech leaves show numerous holes often close together. This is due to the feeding of the adult weevil *Rhynchaenus fagi* (358).

617 Leaf-miner

This characteristic gallery, which

begins as a narrow tunnel and later widens out to become a surface mine is very common and is produced by the larva of *R. fagi* (616).

618 Large holes in leaf
Large holes in beech leaves are likewise very common, and are the result of the feeding activities of the weevil *Phyllobius argentatus* (354).

619 Roll-gall on leaf
Beech leaves are sometimes found to be swollen with their margins infolded. This is due to infestation by the aphid *Phyllaphis fagi*, which congregates on the underside of the leaf and sucks plant juices.

620 Leaf-miner
A common mine located between the lateral veins of the leaf is caused by the caterpillar of the moth *Lithocolletis faginella* (77).

621 Leaf-miner
Several different moth larvae, for example *Nepticula basalella* (74), make mines in beech leaves.

622 Pointed gall on leaf
These common galls, up to 10 mm long, are caused by the gall-gnat *Mikiola fagi*. When ripe they fall off, and may be found in the autumn in quantity on the forest floor. See also 496.

623 Hairy gall on leaf
Galls induced by the gall gnat *Hartigiola annulipes* are cylindrical, hairy structures, 2.5 mm high, at first white and later rusty brown.

624 Vein gall on leaf
The gall-mite *Eriophyes nervisequus* brings about the development of felted ridges along the lateral veins which, at first, are white and later becoming brownish.

OAK

625 Leaf-miner
The gallery begins at the central rib and gradually widens towards the leaf margins. It is caused by the tunnelling larva of the weevil *Rhynchaenus quercus* (359). Pupation takes place in a ball-shaped cocoon in the mine.

626 Folded leaf margin
The edge folded in is always between lateral veins, and curves towards the upper surface. Within the fold there may be one to three larvae of the gall-gnat *Macrodiplosis volvens*.

627 Folded leaf margin
Folds caused by the gall-gnat *M. dryobia* differ from those of 626 in lying at the distal ends of the lateral veins instead of between them. They bend towards the underside. There may be several on a single leaf.

628 Wrinkled leaf-mine
Moth larvae of the genus *Lithocolletis* are responsible for mines which show on the upper surface as slight wrinkling (628a), but are more conspicuous on the underside.

629 Leaf-miner
A mine which does not extend quite to the edge of the leaf and contains reddish brown droppings is made by the larva of a tiny sawfly, *Profenusa pygmaea*. A similar mine which sometimes reaches to the edge is that of the moth larva *Corscium brogniardellum*.

630 Leaf-miner
Burrowing mines are frequent on oak

leaves. These are caused by various species of moth larvae belonging to the genus *Nepticula*.

631 Leaf-miners (Surface miners)
Rounded surface mines, with a silky web enclosed within, are caused by various common species of moth larvae belonging to the genus *Tishceria*.

632 Yellow spots on lamina
These are often accompanied by an upward arching of the surface and are the result of the underside of the leaf being sucked by the aphid *Phylloxera quercus* (44).

633 Pit gall on stem
These are due to the scale insect (coccid) *Asterolecanium variolosum*, the bark forming an elevated ring around each insect, which remains fixed in one place as it sucks juices from the twig. After the coccids have died and fallen off, the twig regains its normal appearance. For other scale insects see 47-49.

634 Gall wasp, *Neuroterus* sp.
About 2 mm long. The female has wings. The life cycle of gall-causing wasps is complex and remarkable. In many species, especially those colonising oak, a bisexual generation of males and females alternates with a unisexual generation consisting exclusively of females. What generally happens is that the males and females of the bisexual phase produce fertilised eggs which are able to survive adverse conditions, such as those prevailing in winter. Development of the eggs is accompanied by the formation of a gall, from which the females of the unisexual generation ultimately emerge. Such females reproduce parthenogenetically, laying unfertilised eggs (generally at

the onset of milder conditions), and these eggs, too, develop in a gall – often one very different in position and appearance from the first. Insects of the bisexual generation arise from these unfertilised eggs. Thus, the life-cycle involves two generations and two different galls.

635 Gall wasp, *Biorhiza pallida*
About 3 mm long. The female has no wings. In this species the male-female generation is winged whereas the parthenogenetic females are wingless.

636 Gall on shoot apex
The point of the shoot expands into an urn-like swelling about 20 mm long. The male-female generation of the gall wasp *Andricus inflator* emerges from this gall in June-July. Its alternate generation develops in small spherical bud galls about 4 mm long.

637 Bark gall
Fluted galls up to 6 mm long disposed in rows along the lower parts of oak trunks and branches. At first the gall is red and very soft but later it becomes brown and hard with a hole in one side made by the wasp eating its way out. It encloses a developing larva which gives rise to the parthenogenetic generation of the gall wasp *A. testaceipes*, the bisexual generation of which originates in oblong galls on the midrib or petiole of the leaf.

638 Bud gall
It is common for some of the buds on young oak in May to develop into red galls up to 7 mm long. In June, the male-female generation of the gall wasp *Trigonaspis megaptera*, comes from this gall. The female generation develops in the kidney-shaped leaf gall shown in 648.

639 Oak apple gall

Up to 40 mm in size. Formed in a bud on older trees, this is at first soft and off-white in colour but later it becomes dry and brown. The gall contains many larvae which develop into the male-female generation of the gall wasp, *Biorhiza pallida*. The adults escape through separate holes in June and July; and the female generation appears from galls up to 10 mm large formed on the roots of the tree.

640 Artichoke gall

Caused by the female generation of the gall wasp *Andricus fecundatrix*, which lays its eggs in a leaf bud. This develops into a gall 15 mm long resembling a small artichoke – at first green and later brown. The solid part of the gall, in between the bud scales, is hard and up to 9 mm long, and falls off late in the summer when ripe. The male-female generation issues from smaller galls in the male catkin of the oak.

641 Oyster gall

Up to 3 mm across, this gall usually lies between the lateral veins on the underside of the leaf. In September or October it falls off and later the same autumn, or in the following spring, the female generation of the gall wasp *A. ostreus* emerges. The male-female generation is developed in small branch galls.

642 Leaf-edge hairy gall

A gall about 2 mm long, which is hairy at first and later becomes smooth. The male-female generation of the gall wasp *Neuroterus albipes* develops in it, and the female generation comes from the very different spangle gall on the leaf-surface (650).

643 Silk-button gall

A round leaf gall about 3 mm in diameter, deepened in the middle and with a thicker edge which is covered by fine hairs (643a). Often present in large numbers on the underside of oak leaves. From this gall comes the parthenogenetic female of the gall wasp *N. numismalis*, and the male-female generation develops in small round leaf galls which project equally from both sides of the leaf.

644 Currant gall

A sphere about 8 mm in diameter resembling a grape which projects from the underside of the leaf. Only slightly protuberant on the upper side (644a). From this gall develops the bisexual generation of the gall wasp *N. quercus-baccarum*, the female generation of which issues from the leaf spangle gall 649.

645 Red pea-gall

The gall is shaped like a slightly flattened sphere, up to 7 mm wide. Often occurs communally underneath the leaf on the veins. Thick-walled (645a). The female generation of the gall wasp *Diplolepis divisa* emerges from it, and the corresponding bisexual generation develops in a conical, hairy gall, 5 mm long, attached to the edge of the leaf.

646 Cherry gall

The gall wasp *D. quercus-folii*, is responsible for a somewhat spongy, spherical gall 20 mm across attached to the veins on the underside of the leaf. The soft wall is very thick (646a). The female generation develops in it, and the bisexual generation in a small purple hairy bud gall found along the length of the trunk.

647 Striped gall
Up to 10 mm in size. Found on the underside of oak leaves. The parthenogenetic generation of the wall wasp *D. longiventris* emerges from it, and the male-female generation from a bud gall similar to that of 646 but greyish geeen in colour.

648 Kidney gall
A somewhat irregular, kidney-shaped gall 2-3 mm long, on the underside of the leaf. In this develops the female generation of the gall wasp *Trigonaspis megaptera*, the bisexual generation of which comes from the gall 638.

649 Common spangle-gall
A gall about 5 mm wide on the underside of the leaf, its convex surface covered by hairs arranged in the shape of a star. Often present in considerable numbers on a single leaf. Falls off the leaf in October and continues to develop on the ground, becoming convex on both surfaces. The gall wasp, which is the female generation of *Neuroterus quercus-baccarum*, emerges in the following spring. The male-female generation comes from the gall 644.

650 Smooth spangle-gall
The female generation of the gall wasp *N. albipes* develops from a flat gall in the shape of a disc of about 3 mm wide, with an irregular margin bent slightly upward. This falls to the ground in October whereupon it swells up appreciably. For the bisexual generation, see 642.

ELM

651 Rolled leaf margin
One side of the leaf is loosely rolled in, whitened and distended. This is due to the aphid *Eriosoma ulmi*, which is heteroecious with red currant whereon it sucks juices from the roots without causing galls.

652 Vein gall on leaf
A common, urn-shaped gall, about 12 mm long, on the upper surface of the leaf which, on maturing in July, develops a split at one end. The causer is the aphid *Tetraneura ulmi*.

653 Leaf-miner
Caused by larvae of the moth genus *Lithocolletis*. A similar mine, which, however, always starts at the middle rib and is transparent, is due to the larva of the leaf wasp *Kaliofenusa ulmi*.

654 Leaf-miner
Due to various moths of the genus *Nepticula*. Those shown here are probably made by *N. ulmivora* and *N. marginicolella*. The latter always follows the margin of the leaf over a considerable length.

655 Brown spot on leaf
These spots, which are frequent on elm leaves, are attributable to the gall mite *Eriophyes filiformis*.

STINGING-NETTLE

656 Galls on nettle
Small white and green swellings, involving both surfaces, are often seen on stinging nettle leaves and sometimes occur also on the stem and inflorescence. They are caused by the gall gnat *Dasyneura urticae*.

ROSE FAMILY

657 Stem galls on raspberry
Round galls measuring up to 30 mm in diameter, containing many larvae of the gall gnat *Lasioptera rubi*, form

irregular bunches on raspberry stalks and, more rarely, on those of dewberry.

658 Stem gall on dewberry
An oblong gall up to 8 cm long in the form of an irregular spindle on dewberry or blackberry stalks. This is at first green but later turns red and finally brown. It contains larvae of the gall wasp *Diastrophus rubi*, each in its own cell.

659 Bedeguar gall on rose
A brightly-coloured gall – green or crimson – of up to 5-6 cm size on a twig or leaf of wild rose. This structure which sometimes consists of several galls merged together, has a firm core but is completely covered with a mossy-looking outer layer of branched filaments. It is caused by the gall wasp *Diplolepis rosae*.

660 Spiked pea-gall on rose
Spherical green or red gall of 5-6 mm diameter with five or six spines up to 4 mm long. Caused by the gall wasp *D. rosarum* on the topside or underside of the leaves.

661 Smooth pea-gall on rose
A spherical gall of 3-5 mm in size on the underside of the leaf. They may also be white or red. Caused by the gall wasp *D. eglanteriae* and, like the foregoing, not rare on wild roses.

662 Oblong leaf gall on rose
A round or, more often, oblong gall, equally visible on both sides of the leaf. Often several grow close together. May be green or red. Due to the gall wasp *D. spinosissimae*.

663 Rolled leaflet margin on rose
Leaflets rolled together towards the midrib occur both on wild and cultivated rose and contain the larvae of the sawfly *Blennocampa pusilla*.

664 Defoliator on rose
May be attributed to the sawfly larva *Cladius pectinicornia*, which eats the rose leaf from the underside.

665 Leaf gall on meadowsweet
Small galls caused by the gall-gnat *Perrisia ulmariae* are often abundant on the leaflets. As development proceeds, the structures pass from green to red.

666 Curled leaves at shoot apex of cherry
Leaves near the apex of the shoots are folded and crumpled and develop irregular protuberances. Caused by the aphid *Myzus cerasi*, the summer generation of which lives on speedwell and bedstraw.

667 Pustulate galls on blackthorn
Small galls, 2 mm wide and 3 mm high, on the leaves of sloe, bullace or plum, are due to the gall-mite *Eriophyes similis*.

668 Rolled leaf margin galls on hawthorn
The edge of the leaf is narrowly rolled downwards and often has hairy spots which are at first white and later brown. Caused by the gall-mite *E. goniothorax*.

669 Leaf rosette on hawthorn
A widespread, abundant gall caused by the gall gnat *Perrisia crataegi*. The growing apex becomes truncated so that the leaves form a rosette. Affected leaves usually bear reddish outgrowths.

670 Rolled leaf margin on crab-apple
Apparently caused by a moth larva (unidentified).

671 Brown spots on crab-apple
Patches on both sides of the leaf, commonly with many small, flaky spots, are due to the gall mite *Eriophyes piri*. Also occurs on cultivated apple and pear trees.

BALSAMS AND LABIATES

672 Leaf-miner on balsam
Leaves of balsam are often mined by a long, irregular gallery which terminates in an equally irregular chamber. This is the result of the feeding activities of a fly larva, *Liriomyza impatientis*.

673 Leaf-miner on labiate
Several round galls are often found together on a single leaf. Produced by the larva of a mining fly, *Phytobia labiatarum*, which may be found on the leaves of various labiate plants.

LIME

674 Globose galls on leaf
Ball-shaped galls formed mainly on the upper surface of the leaf due to the gall gnat *Eriophyes tetrastichus*, which may also cause the leaf margins to be rolled inward.

675 Brown spots on leaf
Another gall-mite, *E. liosoma*, causes rough patches on both the upper and under surface of the leaf. These are white at first, later becoming red and finally brown.

676 Nail galls on leaf
Conical galls up to 15 mm long, often slightly curved, and established on the upper side of the leaf, are due to a third gall-mite *E. tiliae*.

SYCAMORE

677 Leaf-vein galls
Galls up to 4 mm in size, which may be either smooth or hairy on the upper side of the leaf, are due to the gall mite, *E. macrochelus*.

678 Hairy pouch galls on leaf
Hairy spots on the underside of the leaf, which may be seen on the upper side as protruding bumps which are at first clear and later become dark brown. Caused by the gall mite *E. megalonyx*.

679 Red pustulate galls on leaf
Galls measuring not more than 1 mm. Most often red in colour and usually many together on the topside of the leaf, due to the gall gnat *E. macrorhynchus*.

680 Rolled leaf margin
The patch bent over the top side of the leaf is spun firmly down and the cavity occupied by the larva of the weevil *Acleris sponsana*.

ASH

681 Midrib gall
The midrib of the leaf is locally thickened to form a gall extended in length, which has a split in its upper side. This is colonised by the gall gnat *Perrisia fraxini*.

682 Rolled leaf margin
The edge of the leaf is loosely rolled, somewhat swollen and discoloured, a distortion caused by the psyllid *Psyllopsis fraxini*.

683 Leaf-miner

A large mine, at first green and later brown may be inhabited by several larvae of the moth *Gracilaria syringella* (76). The larvae accomplish the early part of their development in the mine, afterwards living communally in the rolled leaves.

684 Leaf-miner

Narrow galleries which expand at the ends made by a moth larva *Praus curticellus* (72) which afterwards lives freely on the leaves, these being lightly spun together.

HONEYSUCKLE

685 Leaf-miner

Very large passages which gradually whiten are made by the burrowing fly larva *Phytagromyza xylostei*. These may also be found on leaves of snow berries.

SPEEDWELL

686 Shoot tip gall on speedwell

The apex of a shoot carries leaves which are folded together with their undersides facing outwards and covered with dense white hair. The causer is a gall gnat *Jaapiella veronicae*.

INDEX

Reference numbers apply to both plates and text.

A

Abax parallelopipedus . 238
Abraxas sylvata .. 95, 192
Acanthocinus aedilis .. 323
Acleris sponsana 680
Acrobasis consociella 78, 186
Adela degeeriella 71
Adelges abietis .. 46, 577
Adopaea lineola...... 178
Aedes 454
Aegeria culiciformis .. 91
Agelastica alni .. 342, 396
Aglais urticae .. 164, 226
Aglia tau 150, 220
Agrilus viridis 271
Agriotes acuminatus .. 275
Agriotes aterrimus .. 274
Agromyza alni-
    betulae 615
Alder Leaf Beetle 342,396
Allacma fusca 26
Allolobophora turgida . 4
Alosterna tabacicolor . 329
Alucita pentadactyla .. 84
Amphipyra
    pyramidea .. 117, 199
Anatis ocellata 298
Anchistrocerus
    parietum 423
Andricus inflator 636
Andricus
    fecundatrix 640
Andricus ostreus 641
Andricus
    testaceipes 637

Anobium punctatum .. 302
Ant Beetle .. 266, 385
Anthophagus
    caraboides 246
Ant-lion 63
Anyphaena
    accentuata .. 507, 523
Apatele psi 110, 197
Apatura iris 163
Aphantopus
    hyperanthus 170
Aphids .. 44, 448, 575,
    576, 577, 578, 601, 602,
    619, 632, 603, 651,
    652, 666
Aphodius fimetarius .. 307
Aphrophora salicis.... 39
Apoda avellana   89, 188
Apoderus coryli .. 362,611
Aradus depressus 58
Araneus
    cucurbitinus   519,528
Araneus
    diadematus .. 518,527
Araneus umbricatus .. 517
Arctia caja ... 133, 205
Argynnis
    euphrosyne .. 169,228
Argynnis paphia 168
Arianta arbustorum .. 570
Arion ater 545
Arion circumscriptus .. 547
Arion rufus 544
Arion subfuscus 546

Armadillidium
    cinereum 13
Army Worm 494
Aromia moschata 319
Ash Bark Beetle 370, 403
Ash Moth 72, 684
Assassin Bug 62
Asterolecanium
    variolosum 633
Atheta fungi 245
Athous
    haemorrhoidalis.... 279
Atolmis rubricollis.... 127
Augiades sylvanus 179
August Thorn Moth 102

B

Bagworms .. 85, 182-184
Banksia tegeocrana .. 536
Bark Beetles .. 364, 365,
    369, 370, 403, 406, 407,
    408, 586
Barred Sallow Moth 121
Bembecia
    hylaeiformis 90
Bena prasinana . 135, 210
Bibio marci 453, 492
Biorhiza pallida   635, 639
Birch Bug 52
Birch Sawfly .. 412, 439
Bird Louse Fly 490
Biston betularia   107, 194
Biting Gnats .. 454, 497
Bitoma crenata 291

171

Black Apollo
  Butterfly 158
Black Arches
  Moth 142, 214
Black Carrion
  Beetle 259
Black-lipped
  Hedge-snail 572
Black Slug 545
Blastophagus
  pinniperda 364, 408, 586
Blennocampa pusilla.. 663
Blow-fly .. 483, 504
Blue Osmia 433
Boarmia repandata .. 106
Boletus Beetle...... 300
Bolitobius lunulatus .. 247
Bombus agrorum 431
Bombus lapidarius .. 428
Bombus pratorum 429
Bombus terrestris 430
Bombylius major 466
Book Lice 27, 28
Bordered White
  Moth 109, 195
Boreus hyemalis 69
Boring Beetles
  300-304, 384, 386
Brachionycha
  sphinx 113
Brachonids 448, 449
Brephos parthenias .. 123
Brimstone
  Butterfly .. 160, 225
Brimstone Moth .. 104
Broad-bordered
  Bee-hawk 157
Brown Slug 546
Buff Arches Moth .. 125
Buff-tailed
  Bumble-bee 430
Buff-tip Moth 137, 211
Bumble-dor 312
Bupalus piniaria 109, 195
Bush Crickets 33, 34, 35

Bush Snail 569
Bythoscopus
  flavicollis 42
Byturus urbanus 287, 382

C

Cabera pusaria 105
Callidium
  violaceum .. 324, 392
Calliphora
  vomitoria 483, 504
Callophrys rubi 173
Calocalpe undulata .. 100
Calocasia coryli 111, 196
Calosoma inquisitor .. 230
Camberwell
  Beauty Butterfly.. 166
Camisia palustris 537
Campaea margaritata 98
Camponotus
  herculeanus 437
Cantharis fusca 262, 377
Cantharis livida...... 263
Capsid Bugs 56-61
Carabus coriaceus 231
Carabus
  violaceus 232, 371
Cardinal Beetle 280, 381
Carrion Beetles 243-260
Carychium
  tridentatum 548
Cassida rubiginosa 347, 397
Catocala fraxini 120
Catocala nupta .. 119, 202
Catops picipes 253
Centipedes 18, 19
Centrotus cornutus 37
Cepaea hortensis 571
Cepaea nemoralis .. 572
Cerambyx scopolii 318
Ceriocera ceratocera .. 479
Cerylon histeroides .. 292
Cetonia aurata .. 313, 390
Chalcoides
  fulvicornis 346

Chelidurella
  acanthopygia 31
Chilosia albitarsis 473
Chimabache fagella .. 73
Chimney-Sweeper
  Moth 93, 189
Chionaspis salicis 49
Chirosia parvicornis .. 574
Chorthippus brunneus.. 32
Chrysobothris
  affinis 270, 378
Chrysomela aenea .. 337
Chrysomela geminata .. 335
Chrysomela populi 336, 395
Chrysopa vulgaris 65
Chrysops pictus 461
Cicindela
  campestris .. 242, 373
Cimbex femorata 412, 439
Cionus
  scrophulariae 360
Cis boleti 300
Cixius nervosus 36
Cladius
  pectinicornia 664
Clausilia bidentata .. 562
Clausilia pumila 563
Clegs 458-461
Clifden Nonpareil
  Moth 120
Clouded Border
  Moth 96
Clouded Magpie
  Moth 92, 192
Clytra
  quadripunctata 334, 394
Clytus arietis 325
Cnaphalodes 576
Coccids 47-49, 633
Coccinella
  decempunctata 294
Coccinella
  septempunctata 297, 389
Coccinella
  quatuordecimguttata 296

Coccinella
quatuordecimpunctata 295
Cochlicopa lubrica .. 551
Cochlodina laminata .. 565
Cochlodina pumila .. 563
Cockchafer .. 315, 389
Cockroaches 29, 30
Coenonympha
pamphilus 172
Coleophora
laricinella........ 184
Columella edentula.... 549

Common Blue
Butterfly .. 177, 229
Common Carder-
bee 431
Common Footman
Moth 128
Common Golden-
eye Lacewing .. 65
Common Hover-fly 471
Common Scorpion
Fly 68
Common Shield-bug 54
Common Vapourer
Moth 141, 213
Common Wasp 426
Common White
Wave Moth 105
Conops
quadrifasciata 469
Conosoma testaceum .. 248
Copper Underwing
Moth 117, 199
Corymbites
sjaelandicus 278
Cossus cossus.... 87, 187
Coxcomb Prominent
Moth 135, 209
Crabro vagus 421
Crab Spider 513
Crambus pratellus 80
Crane-flies 452, 491
Cross Spider .. 518, 527
Cryptococcus fagi 47

Cryptorhynchidius
lapathi 351, 401
Crystal Snail 554
Curculio nucum .. 357, 400
Cychrus scaraboides var.
rostratus 234, 372
Cyclosa conica 520
Cylindroiulus
sylvarum 16

D

Dasychira
pudibunda.... 140, 212
Dasyneura 595, 656
Dasytes caeruleus 267
Dead-leaf Lacewing 64
December Moth .. 146
Deilephila elpenor 154, 223
Dendrobaena arborea.. 6
Dendrobaena octaedra.. 5
Dendroctonus
micans 365, 407
Dendrolimus pini 148, 219
Denticollis linearis 272, 379
Derocrepis rufipes 345
Devil's Coach-horse
Beetle 250
Dexia rustica 489
Diaea dorsata 514
Diaphora mendica .. 132
Diastrophus rubi 658
Digger Wasps.. 421, 423
Dioryctria
abietella 79, 580
Diplolepis divisa 645
Diplolepis
eglanteriae 661
Diplolepis
longiventris 647
Diplolepis
quercus-folii 646
Diplolepis rosae...... 659
Diplolepis rosarum .. 660
Diplolepis
spinosissimae...... 662

Discus rotundatus 553
Dolichopus claviger .. 468
Dolycoris baccarum .. 51
Dorcus
parallelopipedus 311, 387
Dorytomus tortrix 352
Drepana falcataria .. 139
Drepanopteryx
phalaenoides .....·.. 64
Dromius
quadrimaculatus .. 239
Drone-fly 475, 505
Dwarf Millepede .. 17

E

Early Bumble-bee .. 429
Early Thorn
Moth 103, 193
Ectobius lapponicus .. 29
Ectobius lividus...... 30
Elasmucha grisea 52
Elater cinnabarinus .. 276
Elephant Hawk
Moth 154, 223
Empicoris vagabundus 62
Empis tesselata 467
Ena obscura 561
Ennomos
quercinaria 102
Episema
caeruleocephala 116, 201
Erannis
defoliaria 108, 191
Eriocampoides
limacina 443
Ericampoides
annulipes 444
Eriocrania
sparmanella 613
Eriogaster lanestris .. 149
Eriophyes
brevitarsus 610
Eriophyes
diversipunctatus.... 607
Eriophyes filiformis .. 655

Eriophyes
  goniothorax 668
Eriophyes laevis...... 609
Eriophyes liosoma 675
Eriophyes
  macrochelus 677
Eriophyes
  macrorhynchus 679
Eriophyes megalonyx .. 678
Eriophyes
  nervisequus 624
Eriophyes pini 587
Eriophyes piri 671
Eriophyes similis 667
Eriophyes
  tetanothrix........ 597
Eriophyes
  tetrastichus 674
Eriophyes varius 608
Eriothrix
  rufomaculatus 488
Eristalis 475, 505
Ernobius mollis.. 301, 384
Essex Skipper
  Butterfly........ 178
Euconulus fulvus 556
Eucosma tedella...... 579
Eulota fruticum...... 569
Euproctis similis 144, 216
Euura menrinae...... 599
Euvanessa antiopa.... 166
Evertia buoliana 81, 585
Evertia resinella 584
Evesham Moth 101, 190
Eyed Hawk
  Moth 151, 222
Eyed Ladybird 298

F

Fannia 502
Feronia niger........ 235
Festoon Moth 89
Field Grasshopper.. 32
Figure of Eight
  Moth 116, 201

Fir Bark Beetle 366, 404
Flat-back Millepede 15
Flesh-flies 481, 484
Flies 458-490,
  498-505, 574, 615, 685
Forcipomyia 497
Formica rufa........ 436
Fourteen-spot
  Ladybird 295
Fox Moth 147, 218
Fruit-fly 479
Fumea casta 85
Fungus beetles
  .... 285, 293, 300
Fungus gnats
  450, 455, 456, 493, 495
Furniture Beetles 302, 303

G

Galerucella lineola 341
Gall Gnats 582, 588,
  593, 595, 596, 604, 605,
  606, 622, 623, 626, 656,
  657, 665, 669, 681, 686
Gall Mites ... 587, 588,
  597, 607, 608, 609, 610,
  624, 655, 667, 668, 671,
  674, 675, 676, 677, 678,
  679
Gall Wasps ... 634-650,
  658-662
Galumna climata 538
Garden Chafer 306
Garden Tiger
  Moth 133, 205
Gastrodes abietum 57
Geophilomorph
  Centipede 19
Geophilus 19
Geotrupes
  stercorarius 312
Gilema lurideola 128
Gilletteella cooleyi 575
Glanville Fritillary
  Butterfly 167

Glass Snail 555
Glischrochilus var.
  quadripunctatus.... 289
Glomeris marginata .. 14
Glow-worm .. 261, 376
Goat Moth.... 87, 187
Gold Swift Moth .. 88
Gonepteryx rhamni 160, 225
Gracilaria
  syringella 76, 689
Grasshoppers 32-35
Great Green
  Grasshopper 35
Greater Bee-fly 466
Greater
  Horntail 410, 438
Greater Stag
  Beetle 316
Green-bottle Fly .. 480
Green Hairstreak
  Butterfly........ 170
Green Shield-bug .. 56
Green Silver Lines
  Moth 138, 210
Green Tortrix
  Moth 82, 180
Green-veined White
  Butterfly 159, 22
Grey Dagger
  Moth 110, 196
Griposia aprilina 111
Ground Beetles 230-24
Gymnochaeta
  viridis 488
Gyrophaena affinis.... 24

H

Habrosyne derasa 120
Haematopota
  pluvialis 46
Hairy Garden
  Snail 56
Harmandia cavernosa 600
Harmandia globuli .. 600
Hartigiola
  annulipes 620

Harpalus latus 241
Harvestmen .. 529-531
Hazel Snail 567
Hebrew Character
  Moth 114
Helicogonia lapicida .. 568
Helix pomatia 573
Helophilus pendulus .. 472
Hemaris fucifocmis .. 157
Hemerobius humuli .. 66
Heodes phloeas 176
Hepialus hecta 88
Herald Moth 112
Hercules Ant 437
*Hipparchus
  papilionaria* .. 94, 189
Hister striola 260
Hop Lacewing 66
Hornet 427
Hornet Moth 92
Horse-fly.......... 458
Hover-flies 470-478, 503
Hummingbird Hawk
  Moth 156
Hunting Spider 511
Hybomitra collina 459
Hydrotaea irritans.... 487
*Hylastes
  cunicularius* .. 367, 405
*Hylecoetus
  decmestoides* 304
Hylobius abietis 356, 398
Hyloicus pinastri 153, 221
*Hylurgops
  palliatus* 366, 404
*Hypena
  proboscidalis* . 122, 203
Hypogastrura armata.. 20
*Hyponomeuta
  euonymella* 70, 181
*Hyptiotes
  paradoxus* 506, 522

I

Ichneumon.......... 418

*Incurvaria
  koerneriella* 180
Iphigena ventricosa .. 564
Ips typographus.. 368, 409
Isotoma viridis 23
Iteomyia capreae 596
Ixodes ricinus 533

J

Jaapiella veronicae .. 686
Jassus lanio 41
Jumping Plant
  Louse 43
Jumping Spiders 508, 509

L

Lacinius ephippiatus .. 530
Lackey Moth .. 145, 215
Lachnus exsiccator.... 45
Lacon murinus 273
Ladybirds 294-299, 383
Lampyris noctiluca 261, 376
Laothoe populi 152
Laphria ephippium .. 463
Large Coiled Snail .. 564
Large Emerald
  Moth 94, 189
Large Red-belted
  Clearwing Moth 91
Large Red-tailed
  Bumble-bee 428
Large Skipper
  Butterfly........ 179
Large Yellow Under-
  wing Moth .. 118, 198
*Lasiocampa
  quercus* 143, 217
Lasioptera rubi 657
Lasius fuliginosus 435
Leaf Beetles .. 333-347
Ledra aurita 38
Leopard Moth 86
*Leperesinus
  fraxini* 370, 403
Lepidosaphes ulmi 48, 633
Lesser Horntail 411

Lesser Stag-
  beetle 311, 387
Light Emerald Moth 98
Ligidium hypnorum .. 8
Limax marginatus.... 541
Limax maximus 543
Limax tenellus 542
Lime Hawk Moth .. 155
Limenitis sibylla 161
Limenitis populi 162
*Linyphia
  triangularis* .. 521, 526
Liobunum rotundum .. 529
*Liriomyza
  impatientis* 672
Lithobiomorph
  Centipede 18
Lithobius forficatus .. 18
Lithocolletis .. 77, 620, 628
Little Coiled Snail.. 562
Lobster Moth.. 134, 207
Lochmaea capreae 340
Lomaspilis marginata . 96
Long-headed Fly ... 468
Long-horned
  Fungus Gnat 456, 495
Long-horned Moth . 71
Longicorns .. 321-332
*Lophopteryx
  capucina* 135, 209
Lophyrus pini .. 414, 441
Lucanus cervus 310
Lucilia sylvarum 486
Lumbricus castaneus .. 3
Lumbricus rubellus .. 2
Luperus longicornis .. 344
Lyciella rorida 480
Lycosa amentata 510
*Lyda erythroce-
  phala* ..413, 440, 583
Lygus pratensis 61
*Lymantria
  monarcha* 142, 214
Lyonetia clerckella.... 614
Lytta vesicatoria 283

M

Macrocera.......... 456
Macrodiplosis dryobia 627
Macrodiplosis volvens . 626
Macroglossum
   stellatarum 156
Macrothylacia rubi 147, 218
Malachius
   bipustulatus 268
Malacosoma
   neustria....   145, 215
Malthodes
   marginatus 269
Many-plume Moth . 83
Marpissa muscosa.... 508
Marsh-fly 480
Marsh Worm 2
Meconema varium.... 33
Melanotus rufipes 277
Melitaea cinxia...... 167
Melolontha
   melolontha .. 315, 389
Merveille de Jour .. 115
Mesembrina meridiana 482
Mesenchytraeus
   setosus 1
Meta segmentata. 515, 524
Metoecus paradoxus .. 281
Microgaster 449
Microlepidoptera
   70-85, 188, 579, 580,
   585, 613, 614, 620, 621,
   628, 630, 631, 653, 654,
   670, 683, 684
Micrommata
   virescens 512
Mikiola fagi........ 622
Millepedes 14-17
Mimas tiliae........ 155
Mindarus abietinus .. 578
Misumena vatia 513
Mites 534-540
Molorchus minor 330
Mottled Beauty
   Moth 106

Mottled Umber
   Moth 108, 191
Mournful Wasp.... 420
Musk Beetle 319
Muslin Moth 132
Myiatropa florea 474
Mycetophagus
   quadripustulatus .. 293
Mycetophilidae 455
Myrmeleon
   formicarius 63
Myrmica laevinodis .. 434
Myzus cerasi 666

N

Neanura muscorum .. 21
Nebria brevicollis 237
Necrodes litoralis 256
Necrophorus
   humator 254, 374
Necrophorus
   investigator 255
Nemastoma lugubre .. 531
Neobisium muscorum.. 532
Neoitamus
   cyanurus 464, 501
Nephrotoma crocata .. 452
Nepticula .. 621, 630, 654
Neuroterus
   634, 642, 643, 644, 650
Nitidula
   bipunctata........ 288
November Moth .. 97
Nut-tree Tussock
   Moth 111, 196
Nut Weevil.. 357, 400

O

Oak Bark Beetle 363, 402
Oak Eggar Moth 143, 217
Oakleaf Aphid 44
Oak Sawfly........ 444
Octolasium cyaneum 7
Ocypus olens 250
Odezia atrata 93
Oeceoptoma thoracica.. 257

Oedemera femorata .. 282
Oligotrophus
   juniperinus 588
Oniscus asellus 11
Onychiurus armatus .. 22
Operophtera
   brumata 101, 190
Ophion luteus 419
Ophonous seladon 240
Opilo mollis 265
Opisthographtis
   luteolata 104
Oporinia dilutata .. 97
Orange Ant 435
Orange Underwing
   Moth 123
Orchesella flavescens .. 24
Orgyia antiqua .. 141, 213
Oribata geniculata.... 540
Oribatid mite 540
Orneodes hexadactyla . 83
Ornithomyia
   avicularia 490
Orthosia gothica 514
Oryctes
   nasicornis 314, 388
Osmia caerulescens .. 433
Otiorhynchus
   singularis 350
Othius punctulatus.... 252
Oxychilus allvarium.. 558

P

Pale Tussock
   Moth 140, 212
Palomena prasina 50
Panaxia dominula.... 129
Panolis flammea . 126, 200
Panorpa communis.... 68
Paramysia
   oblongoguttata 299
Pararge egeria 171
Parasemia
   plantaginis .. 130, 206

Parasite Flies .. 488, 489
Parasitus 535
Parnassius
    mnemosyne 158
Peach-blossom
    Moth 124, 204
Peacock
Butterfly 156, 227
Pearl-bordered
    Fritillary Butter-
    fly 169, 228
Pear Sawfly 443
Pebble Hook-tip
    Moth 139
Pemphigus bursarius .. 603
Pemphigus filaginis .. 601
Pemphigus
    spirothecae........ 602
Pemphredon lugubris.. 420
Pentatoma rufipes 54
Peppered Moth 107, 194
Perforatella incarnata . 567
Perpolita hammonis .. 559
Perrisia ... 665, 669, 681
Phalera bucephala 137, 211
Pheosia tremula.. 136, 208
Philaenus spumarius .. 40
Philoscia muscorum .. 10
Pholidoptera
    griseoaptera 34
Phosphuga atrata 375
Phragmatobia
    fuliginosa 131
Phthiracarus 539
Phyllaphis fagi 619
Phyllobius argentatus.. 618
Phyllobius calcaratus ..353
Phyllobrotica
    quadrimaculata 343
Phyllodecta
    vulgatissima 339
Phyllopertha horticola . 306
Phylloxera quercus . 44, 632
Phytagromyza 685
Phytobia labiaterum .. 673

Phytodecta viminalis .. 338
Phytocoris populi 59
Phytoecia cylindrica .. 332
Picromerus bidens 55
Pieris napi 159, 224
Pill Millepede 14
Pill Woodlouse 13
Pine Beauty
    Moth 126, 200
Pine Hawk Moth 153, 221
Pine Sawfly ... 414, 441
Pine-tree Lappet
    Moth 148, 219
Pine Weevil ... 355, 399
Pisaura mirabilis 511
Pissodes pini .. 355, 399
Pityogenes
    chalcographus 369
Plant bugs 36-42
Platynus assimilis 236
Platyrhinus
    resinosus 348
Poecilocampa populi .. 146
Pogonocherus hispidus . 331
Pollenia rudis 481
Polydesmus
    complanatus 15
Polyommatus
    icarus 177, 229
Pompilus fuscus...... 422
Pontania capreae 598
Pontania leucaspis.... 589
Pontania pedunculi .. 591
Pontania vesicator 590
Pontania viminalis .. 592
Poplar Hawk Moth . 152
Poplar
    Longicorn .. 317, 391
Porcellio scaber 12
Pot Worm 1
Praon 448
Praus curticellus . 72, 684
Printer 368, 409
Prionus coriarius 316
Profenusa pygmaea .. 629

Proteinus brachypterus. 243
Pseudo-scorpion.... 532
Psithyrus vestalis 432
Psocus nebulosus 27
Psylla alni 43
Psyllids 43, 682
Psyllopsis fraxini 682
Pteronus salicis .. 416, 442
Ptilinus pectinicarnis.. 303
Punctum pygmaeum .. 552
Purple Emperor
    Butterfly 163
Purple Hairstreak
    Butterfly........ 175
Pyrochroa coccinea 280, 381

R.

Raspberry
    Beetle 287, 382
Raspberry
    Clearwing Moth . 90
Red Ant 434
Red-breasted
    Carrion Beetle .. 257
Red Earth-mite 534
Red-necked
    Footman Moth .. 127
Red Slug.......... 544
Red Underwing
    Moth 119, 202
Red Wasp 424
Retinella nitidula 557
Retinella pura 560
Rhabdophaga salicis .. 593
Rhagio scolopaceus 462, 498
Rhagium mordax . 326, 393
Rhagonycha fulva 264
Rhaphidia
    xanthostigma...... 67
Rhingia campestris .. 470
Rhinoceros
    Beetle 314, 388
Rhinosimus planirostris 284
Rhizophagus dispar .. 290
Rhogogaster viridis .. 415

Rhynchaenus fagi   358, 616
Rhynchaenus
   quercus 359, 625
Rhynchites betulae   361, 612
Rhyssa persuasoria   417, 445
Ringlet butterfly   . .   170
Robber-flies   463, 464, 501
Roman Snail   573
Rose Chafer   . .   313, 390
Ruby Tiger Moth . .   131

S

Sabre Wasp   . .   417, 445
St. Mark's Fly .   453, 492
Salticus scenicus   509
Saperda carcharias   317, 391
Saperda populnea.   321, 600
Sarcophaga carnaria . .   484
Saturnid Moth.   150, 220
Sawflies
   412,   416,   439,   442,
   583, 589, 598, 629, 663,
   664
Scallop Shell Moth .   100
Scarlet Tiger
   Moth 129, 205
Sciara 450, 494
Scoliopteryx libatrix   . .   112
*Scolytus
   intricatus* 363, 402
*Scutigerella
   immaculata*   17
Selenia bilunaria.   103, 193
Serica brunnea   305
Sesia apiformis   92
Seven-spot
   Ladybird   . .   297, 383
Sexton Beetles .   254-256
Silpha carinata   259
Silver-washed
   Fritillary Butterfly   168
Single Dotted Wave
   Moth   99
*Sinodendron
   cylindricum*   308

Sirex gigas 410, 438
Sirex noctilis   411
Skipjacks
   270-279,   378,   379
Sloebug   51
Slugs   541-547
Small Copper
   Butterfly   176
Small Eggar Moth. .   149
Small Ermine
   Moth . .   70, 181, 447
Small Heath
   Butterfly   172
Small Tortoiseshell
   Butterfly 164, 226
*Smerinthus
   ocellata* 151, 222
Smooth Coiled Snail   565
Snail Beetle   375
Snails   548-573
Snake-fly   57
Snipe-fly   462, 498, 499
Snout Moth   122
Snow Flea   69
Softwing Beetles 261-269
Soldier Beetles
   262, 263, 377
Spanish Fly Beetle   . .   283
Speckled Wood
   Butterfly   171
Spider-hunting
   Wasp   422
Sprawler Moth   113
Springtails   20-26
Spruce Bark
   Beetle 357, 405
Spruce Weevil   356, 398
*Staphylinus
   brunnipes*   251
Stauropus fagi   134
*Stenocorus
   meridianus*   320
Sterrha dimidiata   99
Stiletto-fly 465, 500
Strangalia melanura . .   328

*Strangalia
   quadrifasciata*   327
*Strophosomus
   melanogrammus*. . . .   349
Stygnocoris rusticus   . .   56
Swallow Prominent
   Moth 136, 208
Syndiplosis petioli   . .   606
Syrphus. 471, 503
Systenocerus caraboides   309

T

Tabanus bovinus   458
Tachyporus obtusus   . .   249
Talaeporia tubulosa . .   182
Ten-spot Ladybird .   294
Tetratoma fungorum . .   285
Tetropium castaneum. .   322
Tettigonia viridissima .   35
*Thanasimus
   formicarius*   266
Thecla w-album   174
*Thecodiplosis
   brachyntera*   582
Thereva 465, 500
Tiger Beetle   . .   242, 373
Tiliacea aurago.   121
Timberman   323
Tipula nebuculosa   451, 491
Tischeria 75, 631
Tomocerus plumbeus . .   25
Tomoxia biguttata 286, 380
Tortoise Beetle.   347, 397
Tortrix viridana .   82, 185
Tree Wasp   425
Trichia hispida.   566
Trichocera hiemalis   . .   457
Trichoniscus pusillus. .   9
Trigonaspis megaptera.   638
Troilus luridus   53
*Trombidium
   holocericeum*   534
*Tryphaena
   pronuba*. 118, 198

V

Vanessa io 156, 227
Vertigo 550
Vespa crabro 427
Vespa rufa 424
Vespa sylvestris 425
Vespa vulgaris 426
Vestal Cuckoo-bee.. 432
Violet Ground
  Beetle 233, 371
Vitraca crystallina.... 554
Vitrina pellucida 555
Volucella bombylans .. 477
Volucella pellucens.... 476

W

Wasp Beetle 325
Wasp Fly 469

Weevils
  348-362, 398, 400, 446,
  581, 611, 616, 617, 618,
  625
White Admiral
  Butterfly 161
White Feather Moth 84
White-letter Hair-
  streak Butterfly .. 174
White-lipped
  Hedge-snail 571
Willow Sawfly . 416, 442
Winter Gnat 457
Winter Moth .. 101, 190
Wireworm 272, 379
Wolf Spider 510
Wood Ant 436
Wood Earwig...... 31
Woodland Snake-
  millepede 16

Woodlice 8-13
Wood Tiger Moth.. 130
Worms............ 1-7

X

*Xylodrepa
  quadripunctata* 258
Xylophaga 499
Xylota segnis........ 478

Y

Yellow Ophion 419
Yellow-tail
  Moth 144, 216

Z

Zebra Spider 509
Zephyrus quercus 175
Zeugophora subspinosa 333
Zeuzera pyrina 86
Zygiella atrica .. 516, 525